I0054139

Reliability Engineering Insights

ASSESS — DESIGN — MANAGE

Arthur Hart

FMS Reliability Publishing
Los Gatos, California

Copyright © 2018 by Arthur Hart.

Licence for Publishing provided to FMS Reliability Publishing

All rights reserved. No part of this publication may be reproduced, distributed or transmitted in any form or by any means, including photocopying, recording, or other electronic or mechanical methods, without the prior written permission of the publisher, except in the case of brief quotations embodied in critical reviews and certain other noncommercial uses permitted by copyright law. For permission requests, write to the publisher, addressed "Attention: Permissions Coordinator," at the address below.

Arthur Hart/FMS Reliability Publishing
17810 Comanche Trail
Los Gatos, CA 95033
www.fmsreliability.com/publishing/

Book Layout ©2015 BookDesignTemplates.com

Book Cover by Hedi LaBel, hedilabel.com

Reliability Engineering Insights / Arthur Hart. —1st ed.
ePub ISBN 978-1-938122-09-5
Softcover ISBN 978-1-938122-08-8

Contents

Dedication

In remembrance of a great manager (Mike Hafner) who inspired high quality in HP products and guided me as I grew in the reliability field.

Forward

T his book provides a valuable window into the role of the contemporary reliability engineer in most current, team-oriented, R&D & manufacturing applications; (both in-house, and out-sourced).

Reliability Engineering (RE) is a combination of technologies and processes that are used to manage reliability-related risks. RE is not just about "does the product work every time you access it": it is also about "does the product function correctly each time over its' expected service environments and lifetime". RE is a composite of field engineering, mathematics, failure analysis, safety, cost analysis, risk management, user training, and various policy issues. Successful RE results in a positive contribution to the products' reception by the customer and the company's reputation in addition to the company's financial bottom lines.

I first met Arthur Hart when he joined a "Think Tank" I was working for back in the late 1970's. We worked together for several years on various projects ranging from radiation electrical effects on both bipolar and MOS Integrated Circuits. This work included 'space war' (EMP) and planet radiation effects on air and spacecraft. Specifically, in Mr. Hart's case, the Jupiter Orbiter Probe was studied. On occasion, we found we could not solve the particular problem at hand, but we could 'set bounds' on the result.

I later joined Mr. Hart at Hewlett Packard where we turned our focus from military to commercial end-user and much larger scale development/manufacturing of products. Responsibilities ranged from engineering, to training and managing engineers, to technical managing of out-sourced areas. Our efforts resulted in numerous patents and a few trade secrets over the next couple of decades. Yes, we actually had fun, winning.

This book captures the culmination of lessons learned along the way. These lessons are by no means unique to where we worked as I discovered when I finished my working career at Xerox.

With that, I encourage you to jump in and keep this close as a reference. Vow to think, make a difference; contribute to the areas you are challenged with.

—John Smyth, Retired Engineer

Preface

After working in failure analysis for several years following graduating in solid state physics, I realized that a much more powerful contribution would be to prevent failures from occurring. The opportunity to join HP was the perfect solution. HP was known for its commitment to quality and reliability in all product areas. This reputation was a primary selling point for its products and was well known over many product types.

The division of HP I joined had the business of developing scientific calculators and personal printers which turned out to be the perfect area to hone my interests in reliability. The management at HP felt so strongly about reliability that they allowed for the development of the many principles discussed in this book. In developing these insights into the 'art of reliability,' a reliability engineer can apply them to the new products being designed making for some fantastic devices. There was a case where HP was so confident in its printers that they referred to them as a tank that you could stand on and not damage it.

As I mastered the art of reliability, I realized that there were many concepts in product reliability that were not in the standard reliability books. I tried to capture these insights in this book so that future efforts in reliability could build on them. These insights create a complete picture about how to use reliability principles.

At the HP-Corvallis site, I became known as 'Mr. Reliability' because of my commitment to these principles and my leadership in forming the Reliability Foundation archive of reliability principles. This foundation was used to train future reliability engineers to carry on the HP tradition.

This book is intended for engineers whose responsibility is the reliability performance of a product. But it could also be used by managers and inexperienced reliability engineers to help build-in reliability while developing quality products. This work stems from my own experiences and knowledge gained while developing and manufacturing of electronic and inkjet products. The driving goal for the book was to make the reliability engineer a significant contributor to the final product performance and highlight the many factors that influence product's reliability.

I would like to thank several of my co-workers (Larry Davisson, Danny Koegler, Paul Watts and Pat Coven) who contributed to the documentation of the tools and coached me along the way on how to apply them. I also want to recognize my managers that challenged me to become part of the team early in the design phase. They also helped me increase my visibility in product teams. Also thanks to Fred Schenkelberg for valuable time and comments in editing this manuscript.

— Arthur Hart

Reliability Engineering Insights

CHAPTER 1

Introduction

W hy are companies 'Recalling' so many products? Why can't they find the problems before customers experience them? These questions are the impetus for writing this book.

Two possible reasons why recalls occur:

1. The development team did not know about product weaknesses before introduction, or

2. The development team DID know about product weaknesses and the company chose to ship the product anyway hoping:

   a. Predictions of failure rates were incorrect;

   b. The actual environment would not be harsh enough to cause failures; or

   c. Failure rates would be acceptable when compared to the added revenue stream produced by beating the competition to market.

When shipments occur with major reliability issues, they may hurt the company's reputation, reduce profits and/or cause high risks to the company's future (especially if customers are physically hurt).

Since recalls are due to a failure of the product to meet the customer expectations, it is the reliability engineer that must champion the customer perspective during the product design. In all cases, the reliability engineer needs to be part of the product development team

1

to understand how reliability can best be optimized for the intended application.

Defining the intended application for each product is critical in order to select the reliability strategy and tools to prevent or minimize recalls. This book is designed to help the reliability engineer get a good start in scoping out the strategy to use in new product development. The reliability engineer can then influence and enhance the reliability decisions made by the development team and connect the needed reliability tools with the specific technology/application.

CHAPTER 2

Basic Understanding of Reliability from a Business Perspective

T his section is a must read for new reliability engineers. It covers the background needed by a reliability engineer to represent the principles of reliability to management and how these principles influence the business model. I have found that management rarely understands that reliability is a combination of design, manufacturing, distribution and customer influences. Poor manufacturing processes or incorrectly defined customer environments can undermine a good design. Selecting the warranty interval is also discussed here since the warranty definition is a powerful concept of reliability and can make or break the profit model for a new product.

2.1 Definition of Reliability

Before discussing how we can select a reliability strategy, it is essential for everyone in the product team to be on the same page as to what reliability means in their particular application.

Reliability is not a term commonly used by the customer. From the customer's perspective, the product will either; correctly perform, not work at all or not work as well over time as the original expectations. From the product development team's perspective, reliability is the ability of the product design to continue meeting customer expectations throughout the life of the product.

The impact of reliability is usually not understood and under-estimated by managers and non-reliability engineers. Good reliability can improve the possibility for future sales, or have a significant effect on sales in follow-on products. There is a general concept that even one bad review of a product is so destructive to a product's quality image that it can affect not only future sales of that one product but also related products from that same company (both now and in the future). High reliability in products creates a valuable brand promise and enhances customer loyalty. A general concept in the reliability area is that one bad review is worth ten good reviews (i.e., A customer is customarily more likely to tell their friends that a product DID NOT meet their expectations than to praise its long-term performance). In the 21st century, we are seeing that social media tools (Facebook, Twitter, etc.) are spreading customer opinions faster and across a wide range of social connections.

One definition of reliability is 'Quality-Over-Time.' A more rigorous definition is 'Reliability is the probability that a component, subassembly, instrument, or system will perform its intended function for a specified time (duration) under specified environmental and use conditions. The key words to remember is 'time.'

If you pull a product out of a box in a brand new state and it fails, the company may categorize this failure as a 'Quality' issue (note; this is a business term since the customer may not understand the difference between quality and reliability). If the product works the first time and then fails in succeeding uses (or over time), the company may define that failure as a 'Reliability' issue (since it is a 'Quality' failure over time).

But if we look at this same product from the manufacturing perspective, the product was tested before leaving the factory and therefore had good quality before shipping to the customer. But by shipping/

handling the product (in warehouses and during store stocking processes), the product fails when the customer opens the package, then from the manufacturer's point of view it failed due to 'Reliability' issues (or 'Quality' over transit time).

Therefore, it is essential for the reliability engineer to represent all areas of a products life and its potential contribution to the failure of the product. Most reliability engineers focus just on the design phase, but even a good design may fail due to:

- inadequate mvanufacturing controls,
- insufficient shipping protection,
- or unexpected disastrous customer use.

An example that I experienced related to customer influences was when we shipped our first version of the inkjet cartridge (used in a printer). The cartridge design included tape over the nozzles to keep the ink from drying out. Unfortunately, the customer did not understand that they must remove the tape when installing a new cartridge and the result was a high failure rates due to the printer 'Not Printing.' In this case, the cartridge worked correctly, but the product failed due to customer influences. Thus no matter how someone labels a failure, it is still a failure.

2.2 Reliability Decisions in Business

It's essential for the reliability engineer to be both focused on the design and the intended business. They need to be involved in all facets of the business that relate to reliability. Below are a few examples of these areas of involvement.

2.2.1 Recalls and impacts to businesses

Good or bad product reliability can impact business success by affecting the volume of inventory needed for support over the life of the product as well as the negative impact on the revenue stream. Reliability estimates influence expected unit replacement inventory, spare parts for service, and turn-around times for repair. These can all affect the bottom line, and development teams rarely consider these and similar issues when introducing a new product.

The decision to ship a product can have long-range cost implications. The following are some recent examples where companies have not done due-diligence in looking for reliability issues before shipping a product:

1. Auto recalls for problems with airbags, breaks, steering mechanisms, emergency braking systems, ignition key retention, etc. (Lawrence 2017),
2. Smartphone battery fires (Mozur 2016),
3. Household problems with washing machines, dishwashers, and ovens (CPSC 2016),
4. Toy problems when used by young children (Safe Kids 2018),
5. Home building products weaknesses with siding, paint with lead, and garage door mechanisms malfunction (NAHB 2018).

These are just some of the well-known examples. Each business category has its unique list of recalls (or known weaknesses that could eventually turn into a recall).

In each of these cases, problems occurred that required a very costly recall (both in revenue and reputation). Most were meant to be precautionary actions to prevent any adverse negative publicity, but some were the result of injury to customers. Depending on the extent of the recall, the cost could be enormous.

Another example that I experienced showing this effect was in the early inkjet printer business. We were developing personal Inkjet products for the office and thus tested our designs throughout the development process so that they would achieve their goals of customer satisfaction in the office. Unfortunately, the prices of personal printers declined significantly, and these same printers sold into the home environment. The unexpected changes caused all sorts of problems and product recalls due to customers using the printers in uncontrolled environments such as high humidity, sunlight, loose pet hair, and mechanical mishandling.

2.2.2 Selecting the optimum warranty periods

A helpful graph (based on an ideal product) shows the total profit and loss over the life of a product from two perspectives; 'revenue' generated from shipping product out the door and 'un-reliability costs' after the customer begins using the product.

Revenue starts to occur when a product ships and customers begin to purchase the product. It includes the purchase price minus the distribution costs, manufacturing costs and incentives made to the retailer.

Un-reliability Cost (negative revenue or loss) occurs from customer call center costs, shipping returns, repair of products under warranty, and the effect on a company's reputation. The actual impact to the company depends on when the failure occurs. If the failure happens after the warranty period, the effects on the total profit are minimal. The largest financial impact is when the failure occurs during the warranty period.

These two primary business drivers (depicted in Figure 1) represent the life of a product revenue stream. We normalized each line to the shipping date of each product (therefore, all products are represented on the same graph so that we can see the effect of a whole revenue stream on one chart.

The goal of any company is to have positive revenue over the life of the product (and as great of profit as possible). The area under each curve is profit or loss (in positive or negative money). The profit is the difference between these two curves (revenue – loss due to un-reliability). Thus both of these curves have to be optimized so that revenue is greater than the cost of un-reliability.

Let's follow each curve from left to right to understand how they interact and affect the bottom line profit.

- The 'revenue' curve (line A) shows money flowing into the company after development and shipments (delayed several months to account for shipping and distribution times). This graph does not show the cost of developing a product which will include the cost of testing and verification done before shipments (that cost

Cumulative Revenue & Costs

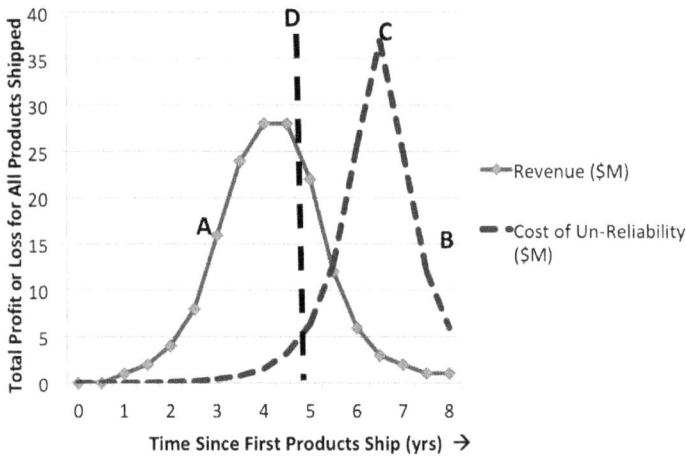

Figure 1 – Product profit and loss when sales revenue flows into the company after shipment and when un-reliability pulls money out.

would show up to the left of the zero axis point). As the start-up ramp ships more products out the door, the 'revenue' curve increases (indicating money starts coming into the company) and then levels off at maturity until product shipments are stopped (note the tail of this line is due to delays in clearing out all products from inventory). Maturity can occur when the technology is no longer competitive or when other companies start taking away some sales.

- The 'cost of un-reliability' curve (line B) is a loss but does not include the cost of setting up a customer service system and failures in the shipping process since we absorb these costs into the cost of development and operation. Eventually, the product will fail to perform as expected. If the customer feels they have received good value for the time used or the unit is out of warranty, they may not return it. The cost to the company, in this case, could still occur from lack of resale or lack of positive word of mouth. Thus there is still a real cost of the field failure even when not returned under warranty (such as a decline in positive word of mouth).

Selecting where warranty would end for products on the time scale (point 'D') can influence the total profit for the company (over the life of the product) through the following factors:

- if the customer does return the product for repair or reimbursement
- how many fail within warranty
- how severe are the failures (e.g., can it be fixed in the field or does it need to be returned to the factory)
- and, whether the warrant is extended.

This will drive the height (point C) and slope of the 'cost of un-reliability' curve (B).

The key for the company is how long to warrant a product to achieve maximum net profit. Good reliability will allow the warranty line (D) to be set longer if that helps sell products. Fixing the product in the field helps to minimize the cost of warranty problems so that curve (C) will be low enough to meet the profit goals of the product. An example of incorporating a good reliability strategy in the product development process is to allow a software problem to be fixed by the customer by downloading an updated software version from a memory stick sent to the customer rather than have the product returned to the manufacturer.

The challenge for each product development team is to estimate where curve (B) falls and push it out as far as possible, make the slope of curve (B) as shallow as possible, and drive the peak value (C) as low as possible. The goal is to achieve a point where the customer feels they have obtained a good value (and therefore decide not to fix or return the product).

Now that the reliability engineer has a common understanding of what is reliability and the factors in optimizing warranty decisions, it is time to discuss the tools to use for achieving the desired reliability.

CHAPTER 3

Factors that Influence the Reliability Approach

Several factors have influenced the approaches used by reliability engineers over the last 80 years. We can summarize these approaches by studying three important categories of influences; Logistics, Technology, and Intended Application.

3.1 Logistics

Some of the factors that have contributed to changes in the logistics of design and manufacturing reliability are:

a. **Sample Size availability** – not only cost but limited parts availability can affect this factor. For example, many modern-day products rely on sub-contractors to deliver samples before integration into the final product, but the timing and expense may push the final integration of parts and sub-systems so close to product introduction that testing time is severely limited. The late arrival of samples leaves little time to analyze the margins and sensitivities of these samples. Often only a few samples (if any) are available for testing purposes across all engineering areas (not just reliability). New inventions may limit the willingness of management to commit these precious resources for testing. Military

11

and space applications are good examples of these. Can you imagine getting even one military airplane or space shuttle for testing to failure?

b. **Schedule shrinkage** – time-to-market is getting more crucial in capturing early adoption customers, thus production may start before all testing is complete. Many new product development times have shrunk from 5-7 years to 18 months.

c. **World-Wide customer diversity** – this may mean different customers have different expectations and environments for use. Customer expectations will drive perceptions of acceptable high quality and reliability. Different ways customers use products may affect the life expectancy.

d. **World-Wide manufacturing sites** – diverse environments and cultures can influence variability and control of materials and processes. The way one operator reads a procedure for module assembly may change its quality and ability to survive the diverse customer environments. Materials, even as simple as water, may have different chemical effects on the product design changing the expected reliability estimates.

e. **Increasing complexity of products** – With the use of modules and high-density microchips, testing each component has become an immense job. More use of modularity assumes the vendor evaluates modules before procurement but how one vendor designs and tests its products before shipment can be widely different.

f. **New technology sub-assemblies** – modules contain many new materials and designs without proven reliability in the field. The subcontractor or system integrator for these sub-assemblies may not have the time or capital to perform all needed reliability studies.

g. **Multiple product applications** – modules and products may be used in many different product designs and applications making for a considerable test matrix. If your product is not the dominant user of the module means the vendor will probably not design it for your application needs.

h. **Spare parts variations** – Over the life of a product, many versions of the same module or different vendors may mean that the module used in service is not the same quality or reliability as that used in the original product design.

With these changes happening all at once in the development environments, reliability engineering approaches must vary between product designs to give the best quality product to the intended customer within the time allowed. It is rare for a new product to consider all of these factors affecting the total life performance.

3.2 Technology

a. The computer – With the invention of high-performance computers and inexpensive microcircuits, statistical sampling and software modeling of data and virtual simulations have enhanced the options for analyzing testing results.

b. Cheaper components – High volumes of low-cost components allow for the use of large sample sizes in tests. More samples reduce the uncertainty of results. Within this category is the use of parts made in low-cost worldwide sites. It may be easier to brute force a study by using large sample sizes in tests rather than model the reliability outcome. Even though the parts are cheaper, the relationship of the initial samples' quality to the high-volume produced parts is unknown.

c. Modularity – More and more parts of a product are being integrated into a module that permits in-the-field replacement rather than a factory repair. The use of modules reduces the cost of servicing the product but changes the cost of failure for any one component. When a resistor fails, the whole module fails, and the cost of replacement is much higher.

d. Statistical tools – Many statistical tools have become available which help the designer and reliability engineer try out new analysis approaches previously overlooked.

e. Testing Approaches – With the invention of sophisticated approaches for reliability testing such as HALT techniques, more accelerated stresses and better estimates of sensitivities to stressors help define the robustness of new designs (Intertek 2018).

3.3 Intended Application

Three major approaches for product reliability have evolved since the beginning of WWII (some 80 years ago);

1. Government Applications (with strict requirements imposing verification of reliability using standard approaches and testing),
2. Built-in Reliability (higher cost products that would be expensive to fix in the field),
3. And more recently, Fix-Problems-in-the-Field (for low cost and high volume products).

Other approaches such as build-test-fix, agile development, and physics of failure studies are techniques embedded in these three general approaches. The reliability engineer now has to select the best method for their application. A discussion of each approach follows:

1. Government Applications – Military and space products fall into this category which has a strict requirement for testing. After WW II, most product development teams (both government and commercial) used these reliability standards trying to make reliability engineering a cookbook approach. Many product types tried to leverage off of these standard approaches.

MIL-STD-883, MIL-STD-785, and MIL-HDBK-217A were created and some adopted for military procurement and commercial products. Weibull.com (2018a) has compiled a set of reliability related military standards. Tables were used to help select the correct sample size for testing. Then it was just a matter of implementing standard tests and estimating reliability levels.

The military approach was modified slightly with the realization that components used by the auto industry could provide higher reliability rather than high priced components that went through extensive testing. The theory here was that the auto industry had such a large customer base, a high volume of products, and a strong desire to reduce warranty costs that they were able to command higher reliability components from vendors that had implemented tighter process control, wider design margins, and better sampling methods. Some military procurements were allowed to substitute auto industry components for select applications (see Hanlon 1996 for summary of justification to use high volume commercial items for military applications).

The business model for development and maintenance may also be different for government contracts. There are cases where military suppliers would make very little profit on the design, yet make significant profit fixing adopted systems during design changes and then profit from maintenance activities.

2. Building-in Reliability (higher reliability expectation products) During the 1980's and into the first decade of 2000, many businesses started trying to build-in reliability rather than do just verifications tests. Many techniques were developed and implemented such as; Total Quality Management, Total Quality Control, and Total Customer Experience. They also applied approaches that tried to expose expected failures and then address them before shipping. Techniques such as; Build-Test-Fix (Schenkelberg 2014), Agile Development (Lifecycle Insights 2018), and Physics of Failure (Pecht 1995), are common approaches in this strategy. These approaches required that the reliability engineer be more involved in the initial phases of a new product and continue throughout its introduction.

As an example; HP was one of those companies that took the building-in reliability approach with inkjet printer products with the result being a widely respected printer for reliability and durability. The concept was that reliability cannot be just a design issue but must encompass all aspects of the product lifecycle (development, manufacturing, supply chain, and customer experience). Ignoring any single area in the reliability model may result in inadequate product reliability. Reliability tools were developed for these new products to have the greatest impact on customer loyalty (i.e., so the customer would perceive a high-quality product). Inkjet printer cartridge development was the primary vehicle for implementing and refining these reliability tools. As inkjet printers became a worldwide business, the team realized that just relying on testing proto products could not achieve the high-reliability expectations.

During early inkjet cartridge development, there were no similar products in the field (other than laser and dot matrix printers), and therefore, little practical learnings could be gained from the field (other than paper dust interactions). There were no established manufacturing sites. Traditional approaches were used at this stage using a single, local proto manufacturing site. The test samples used prototype assembly line units with very highly skilled staff (R&D engineers from the design team and proto-technicians) to train the manufacturing staff. The result was a very controlled product quality giving the design team a good insight into ideal product reliability limits. Once the team resolved all significant issues, the product went into production on the local production line.

With the success of these early products, it became apparent that new manufacturing sites (worldwide) would need to be created to keep up with demand. In the additional manufacturing sites, new factors became apparent;

- local water contamination,
- local vendor variations in materials,
- unskilled production workers,
- shipping of product across large distances,
- documentation interpretation differences (such as during training), and
- New customer expectations from different cultures.

These added factors applied additional stresses to the design, requiring higher margin and additional field repairable requirements. Here are a few examples related to inkjet technology that demonstrate some of the challenges encountered.

Inkjet printers became much cheaper than existing impact printing technology and thus found their way into the personal home market. The development team did not anticipate the influences of the home office environment on failures. All reliability models and testing at that time simulated office/business environments; The home environment included things like uncontrolled temperatures and humidity, dust, animal hairs, and lack of experience by the customer in handling and installing replacement cartridges.

As this new technology became mainstream, the customer expected more features in the ink performance than the original design such as waterproof, smear proof, highlighter proof, and paper independence. These demands forced increased requirements on the chemist formulating inks and on the surrounding materials that interact with the ink. Models increased in complexity. Testing for all possible scenarios became a logistical nightmare.

3. Fix Problems in the Field (typically for lower cost or higher volume products) Recently, time to market (TTM) and complexity have changed the way companies develop products. Any delay doing reliability studies before shipment would possibly keep the new product from achieving the market share goals. Some government contractors also used this approach.

Development cycle times are drastically becoming shorter (Instead of a 5-7 year development cycle, recent evidence shows 18 months to 2-year cycles or shorter). Businesses are now having to deal with constraints on start-up times and simultaneous worldwide distribution. Time to Revenue is now the driving factor.

With the shorter useful life of high tech products (like the iPhone having only 2-3 years before the next generation makes previous models obsolete), fixing the design before it ships and setting up customer service systems is a more reasonable model. By building in extra margin during the design phase, there is less need for testing, and there are fewer problems during product introduction. Unfortunately, modern management does not have the long-range view for reliability but instead is looking at the immediate revenue. They are hoping that the product performs adequately during the short life before the next introduced product. The company saves extra cost and time which equates to Profit. Companies figure they can afford some failures and will build these costs into their profit model. Most modern-day customers understand this and simply want to get the next most recent invention.

In this approach, it is more important for the reliability engineer to consider doing reviews and setting up customer service models to address the expected field failures. Some examples would be to define customer service training documents to allow the customer to reset the original software if it hangs up, or have modules staged around the world for service at regional centers.

Thus the focus of the reliability engineer in this kind of product is to collect knowledge from previous products, conduct disaster verification tests, and use the failures that do occur in customer applications to modify future designs and reliability approaches. Toyota's Prius hybrid car is a great example of this approach. When the car first was introduced, several problems occurred, but after six years of just modifying the design and fixing known problems, the later models were much more reliable (based on my own experience with my 2009 Prius).

CHAPTER 4

Tools for Today's Reliability Challenges

To achieve the required product reliability performance, choose the appropriate approach (government applications, built-in reliability or fix problems in the field). Once this is

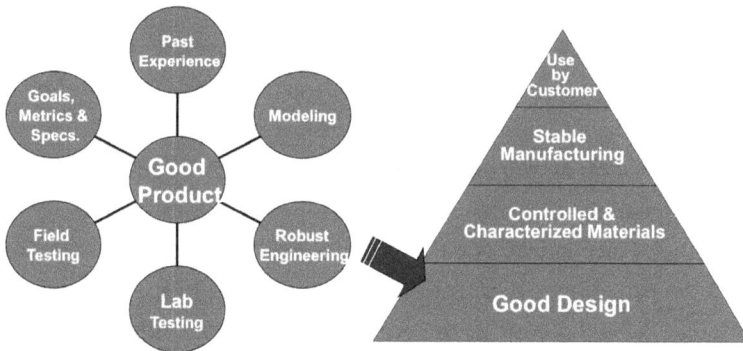

Figure 2 – Engineering areas for a robust product

defined, the reliability engineer selects the tools needed to meet the product goals.

Figure 2 (right side of the pyramid) includes the factors that can be used to influence reliability. Areas to be considered for a new product (bubbles on left side in Figure 2) are all integrated into the design goals.

This map assumes that the design team has time to consider each of these areas. Also, this map assumes that there is some past product experience which may not be correct for start-up products with new technologies. It is the responsibility of the reliability engineer to select the areas to address in the design phase and support services.

21st Century products will require the reliability engineer to create hybrid approaches which combine the best practices of each of the three major types of approaches discussed earlier (government mandates, building-in reliability or fixing problems in the field). The following sections describe reliability tools that may be useful in these approaches.

4.1 Reliability Foundation:

Once the team has selected an approach, the next step is to become familiar with the possible reliability tools for use in a project. These tools will provide the framework for developing a reliability plan.

Below, in figure 3, is a visual map of reliability competencies (also referred as tools or skills) for use in a project. Not all tools will be used in all projects so consider this a map as your toolbox. The toolbox has three types of tools: reliability assessment, design for reliability (DFR), reliability management, and a foundation. Let's name the toolbox the 'Reliability Foundation' (or RFn). Most of these competencies (tools, methods, procedures) are reliability engineering best practices (note; you will find details for each of these skills in the Appendices of this book). The reliability engineer should be able to represent each of these areas when needed in a project.

Reliability Foundation
Core Competencies/Tool/Skills

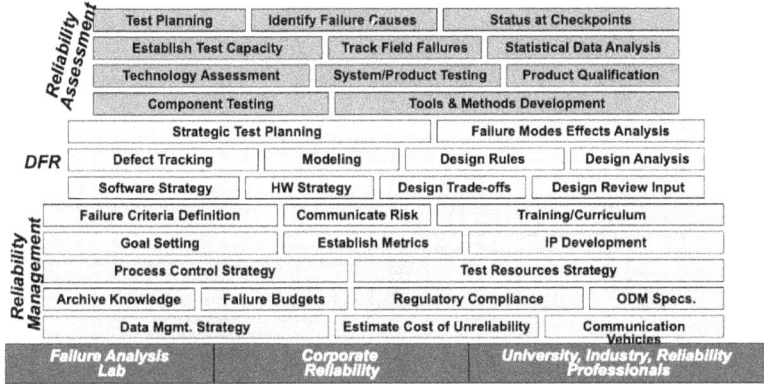

Reliability Assessment			
Test Planning	Identify Failure Causes	Status at Checkpoints	
Establish Test Capacity	Track Field Failures	Statistical Data Analysis	
Technology Assessment	System/Product Testing	Product Qualification	
Component Testing	Tools & Methods Development		

DFR			
Strategic Test Planning		Failure Modes Effects Analysis	
Defect Tracking	Modeling	Design Rules	Design Analysis
Software Strategy	HW Strategy	Design Trade-offs	Design Review Input

Reliability Management			
Failure Criteria Definition	Communicate Risk	Training/Curriculum	
Goal Setting	Establish Metrics	IP Development	
Process Control Strategy		Test Resources Strategy	
Archive Knowledge	Failure Budgets	Regulatory Compliance	ODM Specs.
Data Mgmt. Strategy	Estimate Cost of Unreliability	Communication Vehicles	

Failure Analysis Lab	Corporate Reliability	University, Industry, Reliability Professionals

Figure 3 – Reliability Foundation containing competencies/tools/skills that a reliability engineer may use in different projects.

We can map each of these competencies to a typical product life cycle showing the conventional timing of use. Figure 4 shows the usual timing of each of these tools in a product development lifecycle.

In Figure 4 the high priority tools have a blue font, and regular tools have a black font. You may use each competency independently depending on the application. For example, if a product is the first of its kind, then emphasize DFR tools up front during the early development phases. If the product is a follow-on product and already has some field experience, then using field failure information is more appropriate.

Coupons (vehicles used to explore a sub-assembly or material issue) are also an excellent way to explore potential failure modes without having access to the product. Using coupons allows the experimenter to examine one issue at a time rather than convoluting all failure modes together.

In the early development of inkjet cartridges, little field knowledge was available, so we spent a lot of time on the reliability tools in the Define, Develop phases. Once we were starting to understand the design

Designing For Reliability – When to Use

Blue – High Priority
Black – Necessary

Stages: Concept → Invent → Demonstrate → Characterize → Commercialize

Reliability Core Competencies

Define	Develop	Proto	Start-Up	Volume	Obsolete
Component	Techn. Assessment & Benchmark Project Scoping Functional Arch.	Component Test Tool Plan Coupon Testing for Margins Component HALT and Reviews	Process Change Qualifications Component Test Criteria	Process Audits Design Rules Formed Mfg. Rel. Spec.	Audit improvement (correlate to field)
Sub-System	Research Past Projects Document Needs	Failure Modes Matrix HALT Testing	Learning Cycles Early Proto-Process Defined	Prod.Eng. Partner Post-MR Ownerships Test Capacity Est. Qualifications	Study NTF from Customer Returns
ALL	RACI w/ Partners Historical FMEAs Potential Problem Analysis (PPA)	Test Strategy & Budget Long Lead Tools Goals & Schedule Design Review Build in Reliability Define FA Tools Initial Design FMEA	Test Tool Development Margins Characterized Failure Criteria Defect Tracking System Process FMEA	Mfg. Rel. Spec. Cross-Site Test – Correlation Baselines In-Place	Archive Test Processes & Results Ongoing QOC Support Help Solve Early Fails Rel. Yield Analysis
System		System Test Tool Plan HW/SW/FW Test Strategy	Service/Support Diagnostics RFQ Criteria for Vendors Initiate Rel. Testing	SW/FW Verified ODM Test Correlation Life Data Analysis	Lifetime Buy Impact Support Docs Updated
System Interaction	Model Product Fail Rates Field Data Collection Usage Model Rel. Metrics	System FMEA Reliability Model	Test Criteria – System Correlate Tests w/ Customer Validate Rel. Targets	Document Known Issues Study System Interactions / Desktop Testing	Study Field for Future Projects Validate Rel. Model Validate Usage Model Document New Failures

Figure 4 – Example of a program time line with the reliability tools mapped to show when and where employed (note: this example shows a variety of types of products or sub-assemblies. One should consider using these tools but do not expect to use all of them on each program – only the ones that are appropriate for a specific project).

and its capabilities, then work on tools in the Proto and Start-up areas were applied. During the early development stages, the Functional Architecture, modeling, and test planning were the highest used tools.

Now that the reliability skills are in our toolbox, we can start applying them to a project as needed. Consider the key tool (Functional Architecture) as universal for all projects. Develop a project-specific functional architecture early in the product life cycle. Below we discuss the process with examples of how to use the functional architecture tool effectively.

4.2 Functional Architecture

A powerful tool for both the R&D engineer and the reliability engineer is called the 'Functional Architecture' (or FuncArch). This tool allows all team members designing a product to communicate (in functional terms not physical) across diverse technology areas and to agree on priorities within the system. Since it is in functional language, all parts of the project can visualize their part of the project.

The FuncArch should be employed early in the design phase and championed by the reliability engineer. It is an excellent way for the reliability engineer, design engineers and managers to get to know the design areas and how they depend on each other.

The reliability engineer is probably the best team player to lead this effort because of the system and customer focus they have. This concept is a process to define all functions. While all other engineers are working on the physical layout, the reliability engineer can focus on putting together this map.

Be sure to include both hardware and software functions in this map. Below is an example of the FuncArch concept. Read the map from left to right using inputs (coming in from the left of the drawing) and outputs of functions (emerging from the right of the drawing) with connections to each function inside the product (Note: 'video' is the

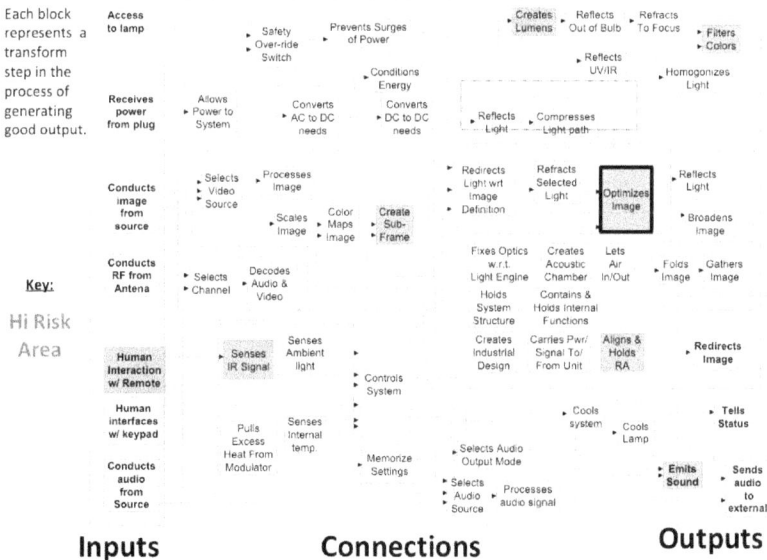

Figure 5 – Functional Architecture map for the video project

product name example in this case). Use this approach at all levels of the product (i.e., system, sub-assembly, component, and software)

The reliability engineer should first interview each project team member in the project and gain their insights into the functions they own. This process is a great way for the reliability engineer to dive deep into the product subassemblies with each designer explaining their function, concerns, and confidence they have in the design. Also discuss known or expected failure modes and how to minimize the chance of failure (i.e., champion the need for more margin, better quality procurement, shielding, etc.). From these discussions, the reliability engineer gets to know the team members and gain an understanding of what areas might be the highest risk.

After individual interviews, the reliability engineer puts together a draft of the entire system FuncArch and presents it to the system design team. This process usually discovers missing functions and linkages that were not understood if functions were looked at independently. The project team should gain consensus on the accuracy of this FuncArch and agree that this map represents all major functions and

Digital Mirror Architecture

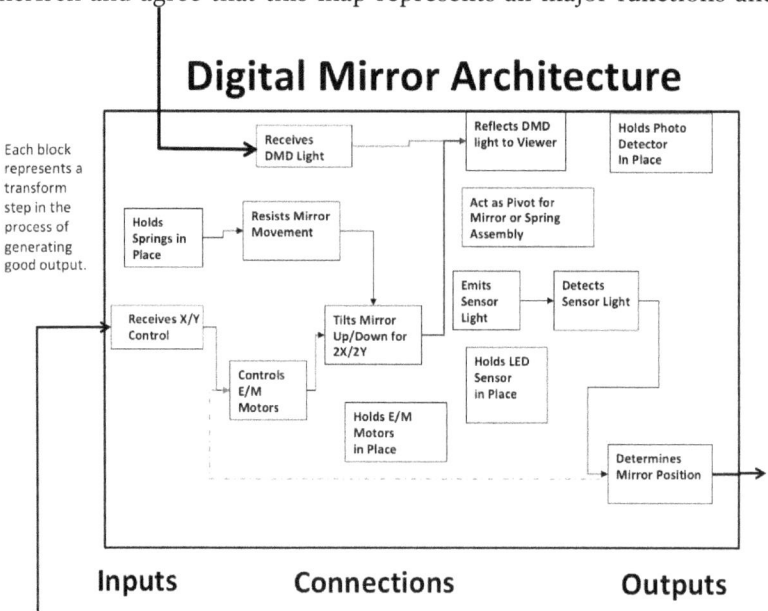

Figure 6 – Sub-System Function Architecture for video project

interconnects. At this point, the reliability engineer is helping the project team understand the viewpoint of the customer.

Each of these system functions can then be broken down into their sub-functions (See Figure 6 for one of the sub-functions of the video project). This process can continue until the reliability engineer understands all basic functions of the product and where possible design weaknesses may occur.

This process provides management with a perspective on risk areas that need to be tracked and verified and also helps define the priority for the effort by the designers and reliability engineer.

Once the team has the first draft of the FuncArch, it can be used in all future management and team reviews to show which functions are still at risk and which ones have reduced their risk or are no longer of concern.

Important: Using the FuncArch process allows the reliability engineer to become a great tool for management to help keep track of the risk of the product during development (and helps keep the reliability engineer visible early in the design phase).

With the FuncArch accepted and the high-risk areas defined, start the follow-on efforts (such as modeling, test planning, literature research and theoretical analysis) targeting those high-risk areas.

4.3 Achieving the Optimum Reliability – Important Concepts

It is also important to consider the entire lifecycle of a product and bring as many unknown factors back into the design phase. Figure 7 shows the kind of factors that you need to consider during the design phase. These are rarely thought about this early in today's product developments due to schedule and cost restraints.

Looking at the reliability action plan from a risk analysis approach gives a holistic view of some of the actions to be done during the development of a new product. Figure 8 shows the critical actions in new product development cycle using risk as the guiding factor. The figure

To Avoid Design Rework, We Need to Think Ahead

In early stages of product development, when options are most wide open, is the greatest opportunity for improving robustness

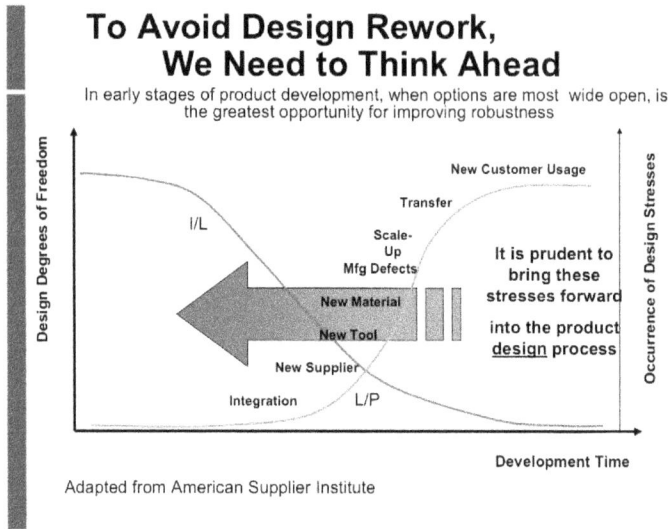

Adapted from American Supplier Institute

Figure 7 – Factors to understand during the design phase.

Reliability Risk Analysis

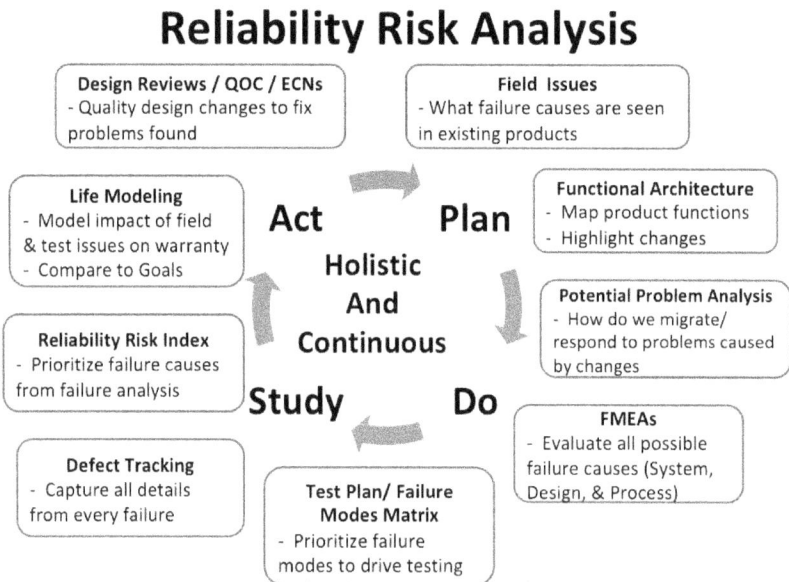

Figure 8 – Risk analysis for showing recommended actions by a reliability engineer in the development of a new product

displays the actions mapped over the typical phases of a project; Plan, Do, Study, and Act.

Let's take a look at two of these areas (test planning and modeling) since they are the most widely used risk analysis tools by reliability engineers.

4.3.1 Reliability Test Planning

Using the FuncArch created in earlier sections, a prioritization of the highest risk product areas can now drive early test work.

Tests to be used can be defined to unmask failures that may occur in the early stages of product life (also called 'Infant Mortality Failures'). Testing for problems in the 'Stable' period (after the early stage and before wear out) and during 'Wear out' period (when defects or weaknesses start causing the product to degrade) are more difficult to measure within the time frame allowed by most new products. Figure 9 defines each of these periods.

The hardest failures to test for are the wear out period failures. Since they take so long to develop, measuring them in the factory before ship-

Bathtub Curve

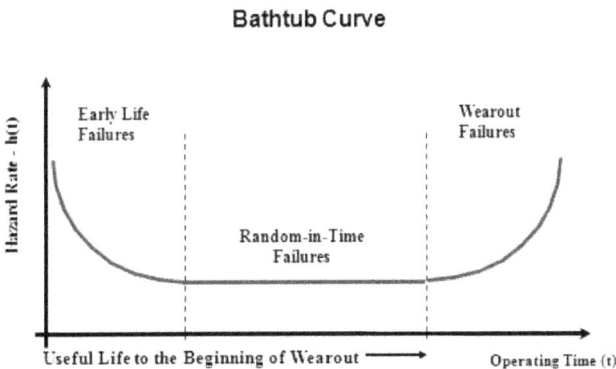

Figure 9 – Definition of Early, Stable and Wear-out types of failures

ping the first products is almost impossible (unless the wear out time is very short – which would be a disaster for the company). Realistically, the best test approach would be to find the infant mortality failures unless you apply very stressful techniques (but then, if you have a failure, it is hard to explain to management why this failure is real).

The following is a draft of testing needed to uncover Infant Mortality, Stable and Wear-out types of failures for our video projector example. Each test is selected depending on the product issues suspected from the design engineer. This test plan includes the kinds of tests used in the video projector project to study suspected risks in this kind of product.

Test Plan Development Example

(Based on Failure Mode, Mechanism Concerns and Models)
Example of Test Plan for the Video project.
- Early Failure Period
  - Tests: Thermal Cycling, Shock, Vibration and Drop Testing
  - Tooling Required: Sub-assembly and Whole system fixtures and in-situ E-tester.
- Stable Failure Period
  - Tests: HALT life tests combining Shock, Vibration, Thermal stresses and Electrical bias stresses.
  - Tooling Required: Device Fixtures, in-situ E-tester, and life test biasing.
- Wear-out Period
  - Tests: Humidity Vibration, Electrical overstress, and Dust generator.
  - Tooling Required: Platform, Plumbing, Electrical, Operation assessment.

Aggressively use early testing combined with a test-to-failure mentality. You want to know how much margin is available for each critical function (not just whether it passes the product specifications). This knowledge will help determine whether manufacturing and material selections and processes are going to be critical stages in the implementation phases.

Two types of test methods found very useful in early testing were HALT and Coupon/sub-assembly testing.

HALT testing (using multiple stresses applied simultaneously and a test-to-failure criterion) is an efficient way to do these kinds of early investigations. The strategy is to find failures that occur when interactions between stresses cause failures earlier than with independent stresses. Each failure should be analyzed (such as by a failure analysis lab) to understand the failure and whether it is relevant to the product usage model.

HALT testing (Intertek 2018) approaches are very valuable ways of finding weaknesses in designs without taking extreme amounts of time or samples.

Here are a few guidelines for the HALT testing approach for discovering hidden defects:

- Every stimulus of potential value is used during the development stage of components and sub-systems to discover their weaknesses
- Stresses are not meant to simulate field environments but instead, they are to find weak links in designs using the fewest tests and shortest test periods
- Stress levels are taken to well beyond normal mission environment until fundamental limits of technology are reached
- To establish the fundamental limits, analysis every early failure even if the failure is found beyond qualification goals

The focus of HALT is on fixing failure modes; not stimulus.

Figure 10 explains the approach used for accelerating and exposing weaknesses in the design using HALT.

HALT Chambers provide quick feedback using combining mechanical/thermal/electrical stresses stimulating field environments to produce failure modes which cannot be detected if tested sequentially (for each stress);

How HALT Finds Useful Life Failures

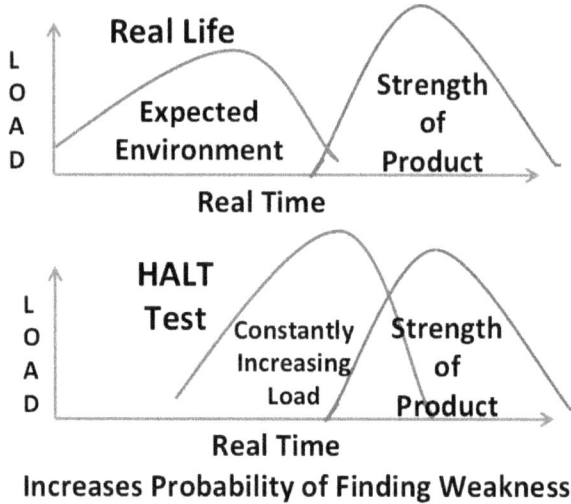

Figure 10 – HALT testing stress vs. margins concept

A good example of this approach used at HP was during the development of a large screen TV with some new screen technology never used previously. We wanted to know early in the development phases if there were any system failure modes unexpected. So we chose the HALT testing approach on the entire TV. To set up the HALT, we placed the assembled TV on a vibration table and connected a heater to the cooling air intake vents of the TV. We positioned sensors and cameras inside the TV to control the heater and internal temperature during operation. While operating the TV and vibrating it, we increased the internal heat until a failure occurred.

The result of this HALT testing on a TV showed many failure modes not found by sub-component manufacturers such as pc board cracking, cable flexing/breaking, and screen cracking. The discovered failure modes would have caused major repair costs if they were not fixed before introduction.

Coupons and sub-assembly testing are good vehicles to use for these early HALT tests since the entire product is probably not available.

Reliability Test Suite
(for Video Product)

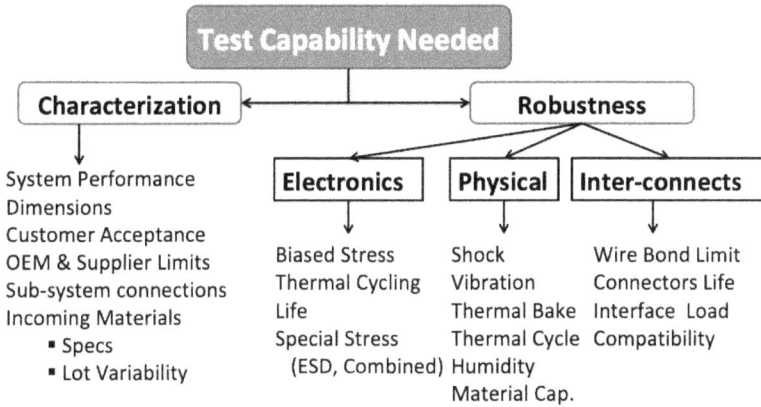

Test Capability Needed

| Characterization | ← → | Robustness |

Characterization

System Performance
Dimensions
Customer Acceptance
OEM & Supplier Limits
Sub-system connections
Incoming Materials
▪ Specs
▪ Lot Variability

Robustness

Electronics	Physical	Inter-connects
Biased Stress	Shock	Wire Bond Limit
Thermal Cycling	Vibration	Connectors Life
Life	Thermal Bake	Interface Load
Special Stress	Thermal Cycle	Compatibility
(ESD, Combined)	Humidity	
	Material Cap.	

Figure 11 – Example of a reliability test plan including both characterization tests (before and after stresses) and robustness tests (to force early failures and degradations).

They also help study suspect areas that are hard to isolate. The coupons study various expected failure modes early in the development phases to understand the mechanisms and causes of the failure. They are also useful when applying very stressful environments impossible on a completed product. Note: a coupon may be an IC chip, material sample, or sub-assembly module but the product is an integration of components, sub-assemblies, plastic, metal, chemicals, and software.

Figure 11 is an example of a test plan developed for the video project discussed earlier. It shows both the characterization tests (done to define if the item has deteriorated or failed) and Robustness testing (the stresses to be applied to force early failures and degradations). These tests helped to understand the margins in each area.

Understanding the variations of the design, materials, and manufacturing will help identify tests needed and interpret failures so that the design can accommodate these variations.

Be sure to track all failures during the product development phases and use these as indicators of weaknesses that remain in the design. Document known weaknesses and include within service manuals for manufacturing and Customer Service notes for use after release to production. Also, show trends of improvements and degradations in some metric so that these can be used to decide to ship or continue optimizing.

Tools to use here are root cause diagrams (Bing Root Cause Diagram 2017), problem solving (Bing Problem Solving 2017) , SQC (*Wikipedia* SPC 2017), & evaluating design margins (Bing Evaluating Design Margins 2017). They help in tracking the issues and put them in perspective when prioritizing their severity and relationship to the real world.

4.3.2 Modeling Reliability

Budgets

To model each major area in the FuncArch for reliability, use past product experience and estimations taking into account:

- the technology used
- materials maturity
- product sub-assembly use
- environments for use
- possible new manufacturing areas
- and prior engineering experience modeling.

Start with the higher risk functions and group the lower risk functions into a few large groups. This model allows for assigning budgets for reliability and giving a large budget for the high-risk functions (Modarres, Kaminskiy, and Krivtsov 1999, 198). Figure 12 shows a typical model for the video project functions discussed in Figure 5.

The model helps establish an understanding between management and the development team related to how individual components/subassemblies will add to the overall reliability of the product and where

Time-Based Reliability Model
How good does a sub-assembly have to be to achieve expectations?

The Time-based reliability of a system for
failures based on component reliability goals.

*Reliability Goals Translate
into a Reliability "Budget" for
each Failure Mode.*

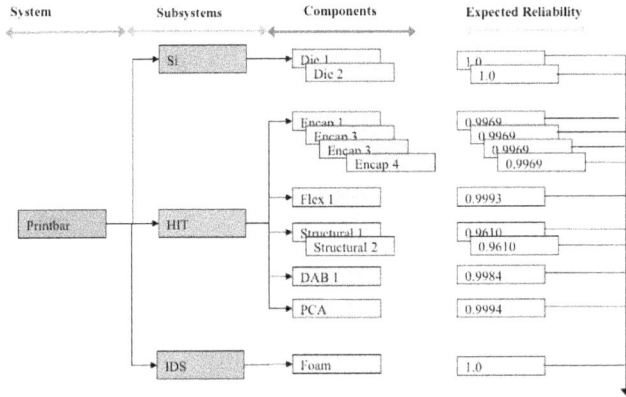

Figure 12 – Reliability model for the
Video Project Functional Architecture blocks

to focus most of the reliability work. Budgets for reliability can then be
established as goals to achieve the necessary product reliability.

Modeling can be difficult if the product uses new technology. If this
is the case, then use the modeling to establish the goals and regularly
compare to early test data (i.e., HALT type tests on prototype and coupon). If the budgets do not seem to match the data, the team needs to
reset the goals by taking some of the higher risk with other parts of the
project.

AFR vs. WRF Use and Interpretation

There are three typical uses of field failure rate models:
1. Cost of Quality: Measure of financial accruals in terms of warranty cost (a good indicator of "monthly bill" to the company
   related to field failures).

2. TCE (Total Customer Experience): Measure success from improvements in product quality and to verify that fixes were successful.

3. Contractual Agreements: Between HP and ODMs/CMs to create expectations of minimum quality levels and ownership.

No single metric can meet the demands of each of these uses. The Cost of Quality typically uses a metric called AFR (Annualized Failure Rate) which is the number of product failures each month divided by the number of units in the field (an accounting metric). TCE is more of a combined engineering and management tool called the WFR (War-

WFR for Bathtub type of Failure Distribution
(assuming 5 month delay between manufacturing and customer usage)

Note how each lot shows bathtub failure rate shape - allows for engineering assessment of lot-to-lot differences.

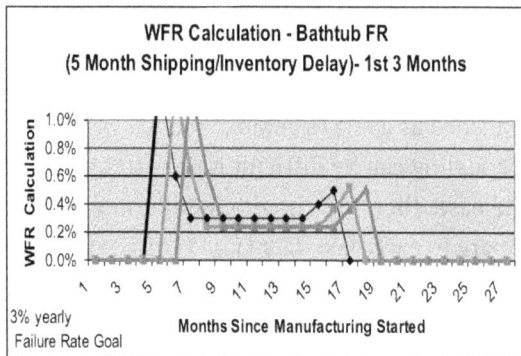

1st 3 months production lots displayed

WFR Calculation - Bathtub FR
(5 Month Shipping/Inventory Delay)- 1st 3 Months

WFR Calculation: 1.0%, 0.8%, 0.6%, 0.4%, 0.2%, 0.0%

Months Since Manufacturing Started

3% yearly Failure Rate Goal

Figure 13 – WFR measurement models for a product during the first year. Each lot is distinguished from other lots so the engineer can define the cause and solution if one is needed. Note how this model can determine the shape of failures and when they occur for each lot.

ranty Failure Rate). Contractual Agreements use both AFR and WFR as part of an agreement which defines who owns the responsibility for each level of problem in both manufacturing tests and customer usage.

The WFR is a measure of each lot separated from other lots by measuring the number of failures (for one lot divided) by the number shipped (for that lot). It is great for engineers and manufacturing managers since it measures each lot separately and allows the definition of early, constant or wear out types of failure modes. The difficulties for this metric are that field metrics do not usually exist to tract this kind of detail. It is also difficult to know how long the product stayed in the supply chain before being delivered to the customer and put into use. Typically, the trackable data is the date code on the package which can be off significantly if a product stays in the supply chain or customer warehouse. Also build-out scenarios will skew this calculation. WFR is usually done during start-up but then loses management support as engineers move onto other projects.

AFR for Bathtub type of Failure Distribution
(assuming 5 month delay between manufacturing and customer usage)

Note how early months lots show no failure rate issues even thought Bathtub failure rate model would indicate early problems.

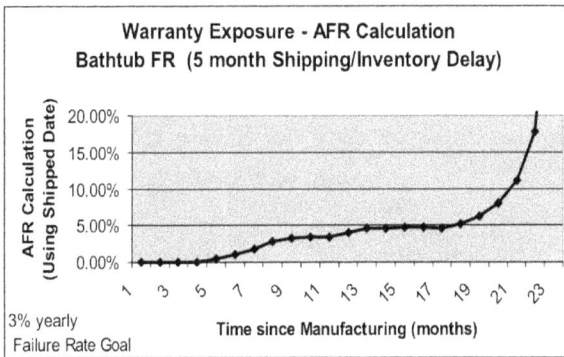

Figure 14 – AFR measurement models for the same product as in Fig. 13 (for the life of production).

The AFR is a good tool for management to measure cost of quality in the field for each product since it is a profit indicator. Thus it has full support in collecting data regularly.

AFR has limitations for engineers in making decisions and determining action points. Since the metric combines the total number failing from all lots produced that are in the field each month and divides by the total units in the field each month, it is not easy to distinguish whether failures were from early lots or a constant failure rate in all lots. It also does not show the shape of the failure curve verses time for each lot and can be totally skewed if he production ramp is putting a high number of units into the field before failures start (i.e. the denominator overpowers the numerator calculation during the early production cycle). Another factor is during the obsolescence stage of production; reliability problems start increasing and the denominator stays constant. When this happens, an artificial spike is seen which can artificially increase management concerns when there are none.

Note that the AFR model (of the same data used in Figure 13) does not detect any issues early in the production cycle due to the increasing number of units shipped during start-up. Thus AFR is not a good tool for engineering to solve problems or make improvements.

4.3.3 Software Reliability Contributions

During the design phase of a product, it is also essential to consider the contributions that software will make on reliability. It is a common misconception that software does not have a reliability factor; only quality matters. This notion is valid only if you consider the initial design of the product.

Reliability is by definition defined as 'Quality over Time.' SW can influence the product reliability after shipment in many ways. The table below shows many software related (good and bad) influences on the actual product reliability field performance. Many of these can induce a failure, but they can also be used to fix failures without returning the product to the factory.

Software Contributions to Hardware Reliability:
- Manufacturing Level
  - Optimization of burn-in life testing parameters and settings

- Coverage and efficiency of end-of-Line diagnostics and margins
- Customer And Field Level
  - Root Cause Tools for Customer Service – using remote connections
  - Techniques for upgrading Firmware in the customer's units at their home (i.e., downloading Firmware updates and fixes using memory sticks sent to the customer)
  - Pro-active warnings of degradations and possible failures to help the customer explain a problem to a service representative
- Data collection
  - Accumulating data on failure rates and types of failures help solve future problems
  - Collection of failure data on components and sub-assemblies is useful information for future products.

If the product fails in the field and the customer can get a quick fix by calling customer service for an upgrade SW, the failure may not be considered major. But if the product has to be returned (or replaced) the customer forgoes a product to use, then in the customer's mind, the failure was major (regardless of what part of the product failed).

For a good example of this approach consider the large screen TV project at HP. During the design phase of the TV, the design team realized that HP did not have a customer service system in place for such a large product. This information drove the design team to try and fix problems in the field. To facilitate this field servicing, the system was designed to indicate a failure mode to the customer and the service technician. The design also allowed for firmware updates by sending memory sticks to the customer enabling for the repair of known problems with the function of the TV. After the TV was released to production, several firmware updates were necessary, and the memory stick solution was cost-effective.

CHAPTER 5

Conclusion: Reliability Engineer's Expectations in Today's Business Environment

The overall theme of this book is for the reliability engineer to be part of the solution in the design phase (not part of the problem). Remember, the reliability engineer is the representative of the customer usage model during development AND implementation.

The reliability engineer should try to be positive and help the design team in finding, solving and interpreting the level of risk in each stage of development.

This engineer should know the kind of product under development (government regulated, high-reliability expectation or high-value low cost) to apply the appropriate level of reliability effort at each stage of the development.

The reliability engineer should also be involved throughout the development cycle and explain the risks at checkpoint meetings and be part of the decision process for risk assessment and resolution. These discussions will also help maintain visibility of the issues at each checkpoint and show the value of having the reliability engineer embedded in the product development process.

From this point forward, a lot of the work is proper planning and implementation of the tools selected.

- Using tools in the Reliability Foundation, drive the design team to build-in margin to account for manufacturing and customer use variability,
- Develop models and tests to evaluate risk areas all the way through the development schedule (not just at the end). Understand each failure and determine whether it will occur in real life applications,
- Set goals for reliability that indicate the risk factors in those functions,
- Test coupons are handy when no other vehicle is available. Test coupons (or sub-assemblies) to failure (not just to the design specification),
- Test proto products knowing that there are some differences between real production units and proto units,
- Evaluate early failures of manufacturing products to verify margin models and expected failure modes. Compare where the failure occurs to the Functional Architecture.
- Evaluate customer returns to determine whether failure modes and rate are in line with early models. Also, compare to the Functional Architecture and budget models.
- Feed manufacturing and customer return data back to next-generation product development team. Second generation products should improve reliability performance by building on these insights.

Details for Each RE Competency (tool) in Reliability Foundation

A. 1 — Defect Tracking

Concept

Collection tool for and documentation of defects as they are experienced, starting at the design phase of a product and continuing through the entire life cycle. For each defect, the following are typically tracked:

- Status
- Severity
- Risk
- Priority
- Owner
- Impact

Defect is defined as any undesired result, or a failure to meet one or more acceptance criterion from our customers, or failure to conform to requirements whether or not those requirements have been specified.

Scope

Is	Is Not
Set of unified defect tracking processes and documents that enables collaboration	Business issues
User interface	Supply Chain non-quality issues
Common dictionary	Marketing tool
Defect life cycle & state model	Finance tool
Administration and support policies	Program management
Training	OEM bug list

Advantages

- Improves visibility of quality issues to all parts of the project team at any one time
- Communicates quality status within and between internal groups
- Provides feedback on quality issues to OEM and vendors
- Helps gain agreement on priority of quality issues and focuses efforts to resolve critical issues
- Tracks history of issues and archives quality issues for future reference

References and Resources:

See Fern (2002) for an introductory article of defect tracking as part of a management process.

Ireson, Coombs, and Moss (1995), Lloyd and Lipow (1984), Klinger, Nadada, and Menendez (1990), and Doty (1989) include outlines and discussion of various defect tracking strategies and techniques.

A. 2 — Design Analysis

Concept

A systematic study of our designs, aimed at maximizing reliability by careful examination of areas such as:

- Design approaches used (e.g., leverage of past designs, complexity, redundancy)
- Materials selection
- Customer use conditions
- Expected manufacturing variability
- Testability, Serviceability
- Expected performance against Reliability targets
- Possible and expected failure mechanisms
- Environmental impact
- Safety

Advantages

- Reduce cost of projects and time to market by identifying areas for improvement as early as possible
- Reduce prototype testing by eliminating obvious failure mechanisms during design reviews
- Prevent recurring mistakes by documenting key learning of each design analyzed
- Provide the means to focus on Reliability concurrently for all design phases and teams

References and Resources

ANSYS (2018) provides a set of design simulation tools including 3D finite element analysis and simulation.

Autodesk (2018) a CAD-embedded finite element analysis software.

Cadence (2018) provides a range of design, analysis, and simulation tools for digital circuits, including Cadence Specman Elite for block, chip, and system level analysis. Dassault Systèmes (2018) a 3D computer aided design modeling software application.

Mentor (2018) provides Vista, among other simulation tools, that is an electronic system level design tool for architectural design exploration, verification and virtual prototyping.

Ireson, Coombs, and Moss (1995), Lloyd and Lipow (1984), Klinger, Nadada, and Menendez (1990), and Doty (1989) include outlines and discussion of various design analysis techniques

---

A. 3 — Design Rules

Concept

A set of basic conditions that all designs should meet – Guiding principles that define a viable design – 'Rules of engagement' for all designers

Define

Industrial designs are developed under a set of complete rules and procedures that define how the design is generated, archived, verified, communicated and implemented. These rules are defined by each design organization specifically to meet the characteristics of the products or services developed. They all, however, try to steer designers to create solutions that have some basic characteristics:
- Are simple. The best design is always the simplest solution
- Are innovative and appealing solutions that consistently perform the required function
- Minimize part count and cost
- Are easy to produce
- Are easy to maintain
- Are easy to assemble
- Are easy to test
- Are environmentally friendly

- Utilize existing technologies (when feasible)
- Use standard components and materials (when available)
- Are thoroughly documented

References and Resources

See Poll (2001) for an overview of design for manufacturing design rules.

Davey (2003) examines the design principles in select Japanese products.

See Blankenhorn (1996) for an examination of surface mount technology design rules.

Ireson, Coombs, and Moss (1995), Klinger, Nadada, and Menendez (1990), and Doty (1989) include best practices and common approaches to developing and using design rules.

A. 4 — Design Trade-offs

Concept

An exchange of one design feature in return for another, specially relinquishment of one feature for another regarding as 'more desirable'

Shall I use the net or the rod today?

Do I want to catch the big one or catch many small ones?

Define

The need to trade one feature for another 'more desirable one' is a constant of every design activity.

When we need to create a mechanically robust component, for example, we have to 'balance' other features like weight and cost which are inherent detractors to robustness. In electrical design it is common to trade speed of transmission of data and signals with power consumption and heat generation or radiated emissions.

It does not matter if we are designing software, hardware or services, one of the key elements of the design effort is in the establishment of priority in the features of the product and the adequate tradeoffs made during the process.

To help establish an adequate 'tradeoff policy,' it is recommended that the designer:

- List all product features
- Assign a weight or priority rating to each based on the final user perspective. Ask: What is more important for my customer?
- Review, at the end of each design phase, that all design tradeoffs have been made supporting the priorities or weights assigned

References and Resources

See Freemantle (2002) for an excellent treatment of decision making.

Ashby (2017) examines the process of selecting the right materials for your mechanical design.

Ireson, Coombs, and Moss (1995), Klinger, Nadada, and Menendez (1990), and Doty (1989) include framing discussions of design tradeoff practises.

A. 5 — Failure Modes and Effects Analysis (FMEA) Also Potential Problem Analysis (PPA)

Concept

A forward-looking, structured method to consider each potential failure mode of a system, design, or process; estimate frequency and detectability, and to ascertain the effects on the customer.

> Note: Failure Mode is the manner in which a component, subsystem, or system could potentially fail to meet the design intent, performance requirements, or even customer expectations.

Scope

Is	*Is Not*
Structured method	Absolute equation
Way to link potential failure modes with specific system elements	Policy setting method
Product development tool in early stages	Facilitating by a reliability expert
Collaborative effort with select team of experts	Detailed course or instructions
Forward looking	Substitute for engineering knowledge
Prioritization of risk areas	Failure analysis or Root Cause Analysis
Started early in development cycle	Design Validation

Advantages

Failure Mode and Effects Analysis (FMEA) – A FMEA is a systemized group of activities intended to:

- Recognize and evaluate the potential failure of a product/process and its effects;
- Identify actions which could eliminate or reduce the chance of the potential failure;
- Document the process. Identifies potential systemic, design-related or process-level failure modes
- Identifies key critical and significant characteristics of a design
- Provides an organized, systematic approach to consider and reduce product failure risks
- Provides a means to compare competing designs
- Is an input into design validation and product test plans
- Promotes collaboration across team

- Organizes existing knowledge to identify gaps
- Promotes focus on highest priority risks
- Reduces opportunity of releasing with same failure

References and Resources:

The Annual Reliability and Maintainability Symposium, RAMS, generally has one or more paper sessions devoted to FMEA applications and advances. See Onodera (1997) for one such conference paper.

Whitcomb and Rioux (1994) provide an example of an FMEA in the semiconductor industry.

Automotive Industry Action Group, AIAG (2008), has a short book detailing how to conduct an FMEA analysis. It is written for the auto industry, yet the process is appropriate for the electronics industry.

A. 6 — Hardware Strategy

Concept

A systematic plan of action focused on the design, supply chain, inventory, distribution, maintenance and disposition of hardware that consistently meets customer expectations while maximizing profits for the organization.

Define

Each organization in the company has a specific strategy to meet the requirements of the market it serves. They all however, steer to common targets regarding their hardware. Examples include:

> "Reliable innovation at a price our customers can afford, delivered with an experience that set us apart. We deliver high tech, low cost, best customer experience"

- Is highly reliable. Warranty returns undermine company profits and prevent future sales.

- Is innovative and appealing, setting new trends in the markets we play.
- Minimizes part count and cost.
- Is environmentally friendly and endorsed by local safety agencies in all geographies where it is sold.

Reliability Note

There are BASIC hardware design rules that are part of a comprehensive Hardware Reliability Strategy. Examples are

- Selecting only quality components from reliable and qualified vendors
- Isolating sensitive components
- Creating redundancies to reinforce sensitive components
- Definition of a service plan if applicable (if the product is going to be serviced, it has to comply with design for serviceability conditions)
- Defining Reliability budgets (quotas) for each component or subassembly in support of an overall product reliability goal

References and Resources:

See Anthony and Govindarajan (2003) for a focus on management control systems.

Panagacos (2018) discusses business process management.

Scheiber (2001) examines methodologies to successfully create a circuit board test strategy.

See Cohen (1995) for a detailed discussion on quality functional deployment use.

Ireson, Coombs, and Moss (1995), Klinger, Nadada, and Menendez (1990), and Doty (1989) generally discuss hardware reliability approaches.

A. 7 — Modeling

Concept

The process of representing in detail, a system malfunction or a problem (under controlled conditions), to be used for testing, trouble-shooting, or further investigate characterization.

Define

Modeling requires a clear understanding of what characteristics are needed to analyze as is normally used when it is not feasible to use the final or complete product or process. Once it is known what character-istics have to be duplicated for analysis, then a decision can be made on how to recreate these characteristics. Models can be made math-ematically (using computer generated models) or physically by using prototypes for testing.

The key to successful modeling is to recreate the characteristics for analysis as close to reality as possible.

Processes (manufacturing, engineering or business) are especially suitable for computer modeling. This allows for early troubleshooting and fine tuning prior to real implementation avoiding costly and timely issues.

References and Resources

Winston and Albright (1997) provide approaches to modeling using a spreadsheet.

Fowler (1997) provides a detailed treatment of applied mathemati-cal models.

A. 8 — Checkpoint Status

Concept

The basic tool to monitor and report progress of Reliability activities. Normally coded by colors:

- Green – Completed
- Yellow – In progress with no significant risk to fail
- Red – Failed or imminent risk of failure.

Define

All Reliability activities must include scheduled checkpoints to assess and report progress against objectives.

New Product Introductions (as one type of many different Reliability activities) need a consistent, predetermined calendar of checkpoints. These checkpoints are used to "sign off" the different phases once the predetermined goals are achieved.

Examples

Design Verification Testing (DVT) is concluded only when the status is green on all failures identified and a mean time before failure (MTBF) of XXX thousand hours is achieved with an YY percent of confidence.

Product is released for volume production only when the status is green (no critical issues present) after product qualification testing.

> "Product reliability is an important discriminator in today's global marketplace but does your organization know how its doing compared to competitors in terms of designing and building reliability into its products? Could/should you be doing more "upfront" reliability activities to improve customer satisfaction, to reduce warranty costs, and to reduce in-plant rework costs? Do you even know how your products are doing in field operation? "
> – Qualitron, training material

Resources and References

See Burgelman, Wheelwright, and Maidique (2000) for a complete treatment of strategically managing technology and innovation.

Ulrich and Eppinger (2004) focus on the product design and development stages of the product lifecycle

See Ziglar (2002) for a classic treatment of goal setting.

Ireson, Coombs, and Moss (1995), Lloyd and Lipow (1984), Klinger, Nadada, and Menendez (1990), and Doty (1989) each connect reliability to the product lifecycle stage gate checkpoints.

A. 9 — Component Testing

Concept

One of the most important parts of the development process, where various components are tested over a wide range of conditions to insure the best alternatives are chosen for the expected level of performance of the system.

Note: Component: a constituent element of a system

Scope

Is	*Is Not*
Designed to identify the strong and weak elements of the whole system. The product's final reliability is dependent on the weakest element.	Simple coupon tests.
The basis for a reliability growth plan.	System testing.
Recommended procedures to ensure test rigor.	System interaction tests.

Is	*Is Not*
Intimately linked to several other reliability tools.	
Supported by test tools developed for component testing	

Advantages

- Will help identify those components that require reliability improvements to meet the system performance expectations and margins as well as components that represent opportunities for cost reduction.
- Can be done before the entire system is mature enough to test.
- Can be done when the system is too large or expensive for significant system testing.
- May be able to test a component in a way that it could not be tested once part of the system.
- Useful when the component is difficult to access once it is part of the system.
- Can reduce both cost and time to market by finding problems earlier.

References and Resources

Wallnau (2004) outlines the Carnegie Mellon, Software Engineering Institute software component certification process.

HBM Prenscia Inc. (2018) specifically the ReliaSoft team provides a comprehensive article on design for reliability including component level analysis.

ABS Group (2018) provides seminars on a range of topics, including quantitative analysis techniques suitable for component modeling.

ReliaSoft (2018a) within their ReliaWiki site provides an article on evaluating component reliability importance.

Ireson, Coombs, and Moss (1995), Lloyd and Lipow (1984), Klinger, Nadada, and Menendez (1990), and Doty (1989) include a range of discussion concerning component testing.

A. 10 — Field Service and Support

Concept

Reliability implications in customer environment and how service of products in the field is applied.

Discussion

When you buy a new PC, camera, printer, or TV, you want a dependable device from a manufacturer that's committed to supporting its customers.

Every year, consumers purchase millions of computers and peripherals. And every year, millions of those devices break down.

For anyone who plans to buy a piece of hardware, the overall reliability of a vendor's products and the quality of its service are important considerations. Unfortunately, much of the information that people use in deciding which product to buy is fragmentary and anecdotal—not the kind of data you'd want to base a three – or four-figure decision on.

Perhaps the most surprising thing about our most recent survey results is how closely consumer opinions about reliability and service compared to those reported in previous years.

> "We're trying to move to a leadership position in service and support, and that's taking a large investment and some time."
> — Jodi Schilling, Vice President of HP American Customer Support Operations

Schilling and Brent Potts, vice president of the Web support operation, say that the company is focusing on three key areas: the initial design of its products, the products' operational performance and reliability, and the way the company supports its products. The last of those seems to be getting most of the attention: Because of this, companies can ramp up online FAQ archive, radically expand its forum-based support (where experts and users can get together to talk shop), introduce video-based tutorials, and built new programs such as what is called Ambassadors around a team of 50 experts who reach out directly to more-vocal customers (read: major bloggers) to help solve problems.

Hiring additional support staff is usually more expensive and requires extensive training (people and materials) and is hard to obtain consistency. Readers continue to complain about communication difficulties with overseas support reps and about the poor training that some tech staffers, whether foreign or domestic, seem to have received.

Service Measures

Phone hold time

Based on the average time a product's owners waited on hold to speak to a phone support representative.

Average phone service rating

Based on a cumulative score derived from product owners' ratings of several aspects of their experience in phoning the company's technical support service. Among the factors considered were whether the information was easy to understand, and whether the support rep spoke clearly and knowledgeably.

Problem was never resolved

Based on the percentage of survey respondents who said the problem remained after they contacted the company's support service.

A. 11 — Materials Selection and Compatibility

Concept

Since materials are foundational elements of any reliability study, knowing the capability and limitations of product materials allow for modeling performance in real life and building-in margins to compensate for production process variations, environmental applications, and compatibility with other product materials. Previous testing and history can be used to evaluate whether materials will have sufficient margins but in new products and applications, this may not be available requiring your own testing schemes.

Benefits

- Design: Designing with full understanding of the limits and compatibility of the materials being used allow the designer to make the appropriate trade-offs when optimizing performance, costs and time to market. Sometimes the best material is one that has been fully characterized rather than choosing the latest and greatest material that just arrived on the market.
- Production: It is important to know the requirements and capability of materials used during the assembly of products so that they can be introduced into the processes without changing the performance and reliability of the product. Having established tests and criteria for all products helps maintain performance over the life of production. Many times production processes are changed or moved to another site. When this happens, knowing what has been changed and how it will affect the performance of the product is critical in making a smooth transition to the future processes.

Discussion

Examples of material and compatibility areas:

- In Ink-Jet printing, development is done in one lab and in one production facility. But as production ramps, more production sites need to be brought on-line. These new lines change the local vendors for materials and different customs of training that can affect the compatibility of materials in a product.
- Materials may be deleteriously affected by UV-B radiation exposure. Typical materials include man-made polymers (plastics and rubber materials) as well as naturally-occurring biopolymers such as wood, wool skin proteins, and hair.
- Both natural and man-made materials are widely used in the building industry.
- Some of the building products, for instance roofing materials, plastic or wood siding, window frames, glazing, water tanks, rain ware (gutters) and plastic conduit are routinely tested for sensitivity to sunlight during regular use.
- Numerous plastic products exposed outdoors photodegrade slowly on exposure to solar radiation, steadily losing their useful physical and mechanical attributes.
- This degradation is accelerated by higher ambient temperatures and in some polymers by the presence of moisture and pollutants in the atmosphere.
- The use of plastics outdoors is feasible only because of the use of efficient photo stabilizers designed to slow down this degradation process. A large body of information exists on stabilizer technologies aimed at extending the service life of plastics exposed to outdoor environments. In regions where climatic changes (caused by global warming) result in higher rainfall or humidity, the situation is even further exacerbated. In both wood and moisture-susceptible-plastics, photo-degradation can be significantly accelerated by the presence of high humidity, particularly at the higher temperatures. The effect is well documented and higher temperatures and humidity are routinely used to accelerate the photo degradation of plastics and rubber in laboratory exposure tests.

Resources and References

See the article by Andrady and Hamid (2003) for a discussion on the effects of climate change on materials.

A. 12 — Procurement

Concept

Procurement practices depend on many factors depending on what is procured, who it is procured from, the intended customer and the government that controls the requirements on distribution. It is important to understand these boundary conditions and document them for future reference.

Procurement practices must be adhered to for both legally and reliability reasons in order to achieve consistency across time, product lines, and countries.

Define

Examples are:
- The development and production of ink jet cartridges. When the first one was developed, it used one production line in one country and with one set or materials. As quantities increased and production lines expanded in many countries, local materials, local staff and cloned production machines affected the interaction chemistry in the cartridges which affected the color gamut, life of the ink and stability of the product.
- Cost Reductions are important considerations in long term reliability of products. Saving money sometimes results in getting cheaper parts that do not have the intended result such as getting a part from another vendor that leaves out crucial elements or lacks the strength characteristics. Costs must be traded off

with the testing to prove that the change is sufficient and meets the original goals.

- Auditing for reliability is a very difficult function due to the time for testing and sample size. Using HALT and HASS type strategies allows for verification that margins are not lost when changes are made in procuring materials and parts. Convincing management that a change is not worth the risk is difficult because of the limitations on samples and time to do tests.

- Procuring sub-systems from another vendor that has a proven reliability record is a good way to leverage a products value and reduce the time to market. But the control of reliability over the long run is more difficult due to the lack of visibility into the vendors practices.

Resources and References

Blanks (1992) write about how to achieve product reliability objectives via your procurement practices throughout the product lifecycle.

A. 13 — Product Qualification

Concept

The final phase of the Product Development Cycle. Aimed to confirm the product meets all functional requirements as defined in the product specification and therefore is released for volume production.

Areas that require Qualification are: Any changes in process, material, equipment, or manufacturing site made during the life of the product so that quality/reliability is maintained at required levels."

Define

Product Qualification consists of a number of tests that reflect each and every one of the performance parameters identified in the product specification.

These tests are defined in a Product Qualification Plan that is written beforehand as part of the initial development documentation phase and includes an explicit definition of failure, description of the specific conditions for each test along with sample sizes to use on each experiment.

This Product Qualification Plan along with the reports for each test included constitutes documentation that supports the decision to produce in volume and should be considered and archived as official product documentation.

Typical Qualification Test Areas – Hardware Products

Environmental

- Operational Temperature and Humidity
- Storage Temperature and Humidity
- Operational Vibration
- Operational shock
- Transportation Vibration (Packaged)
- Transportation shock (Packaged drop)
- Operational dust
- UV exposure (for plastic enclosure)
- High and low altitude Operational
- Contrast and readability of LCD display before and after T/H
- Contrast and Readability of LCD display before and after Electrical Life
- Operational Noise levels generated

Mechanical

- Number of printed pages
- Number of paper jams
- Number of scanned images
- Life of doors (number of open / close cycles)
- Life of trays (number of open / close cycles)
- Life of lids (number of open / close cycles)
- Life of keys
- Life of switch (or switches)

Electrical

- Electrical life (on all electronics)
- Voltage stress (high and low)
- Power cycle (number of on / off cycles)
- EMI (Electro Magnetic Interference per FCC class B???)
- EFT (Electrical Fast Transient)
- ESD (Electro Static Discharge)

Human Interface

- Inspection for sharp edges on all accessible areas (including maintenance areas)
- Safety levels of fixer fumes during operation in low ventilation conditions
- Chemical reaction to cleaning agents on enclosure and document glass
- Sweat on keys, touch pads and document glass
- Overforce (abuse) on keys, touch pads and document glass
- Spills (water, coffee, soda)
- Mechanical abuse while clearing a paper jam
- Force needed to open access to maintenance areas

Typical Qualification Test Areas – Software Products

- Configuration
- Installation
- Compatibility and interoperability
- Documentation and help files
- Fault recovery (Corrupted or missing files, power loss, accidental shut down, etc.)
- Performance (user and system response times, external interface response times, CPU use, memory utilization)
- Security
- Stress testing (peak loads)
- Usability
- Volume (of data or concurrent users)

Resources and References:

See Willborn and Cheng (1994) for a treatment of quality assurance concerning product qualification.

Ulrich and Eppinger (2004) focus on product qualification during the product development process.

Ireson, Coombs, and Moss (1995), Lloyd and Lipow (1984), Klinger, Nadada, and Menendez (1990), and Doty (1989) include a range of discussion concerning product qualification.

A. 14 — Product Testing

Concept

The design, implementation and analysis of experiments aimed to demonstrate the reliability of a product or system. These experiments simulate conditions (stresses) that the product will experience during the intended use and can include some form of "acceleration" to expedite results.

Product

Commodities offered for sale; an artifact that is the result of a process; anything that is the result of generation, growth, labor, or thought, or by the operation of involuntary causes; as products of manufacturers; the products of the brain

Tool and Method Development:

- Product level test
- A design verification, production release, and product quality audit method
- Margin testing (using approaches such as HALT and HASS)
- Allows engineering groups to do Pareto charts of failure modes.
- Way to discover new or unexpected failure modes
- Warranty estimation tool
- All possible tests for all situations

Advantages

- Identifies potential design-related, process-level, or materials—limits product weaknesses and failure modes
- Provides a means to compare competing designs and vendors or products
- Can be a design, process control or audit tool
- Provides the only vehicle to test integration issues
- Vehicle to obtain the customer perspective on product performance and limitations.
- Provides management with a final assessment on product performance and limitations.
- Identify possible cost reduction areas
- Quantify and improve field reliability
- Improve customer satisfaction
- Increase market share
- Improve profitability
- Meet the Reliability and Quality goals

- Verification of compliance to an environmental class
- Identification of design defects and failure modes
- Identification of wear out mechanisms
- Identification of process defects in production
- Identification of latent defects (not discernible until after the product has completed the manufacturing process.

A. 15 — Technology Assessments

Concept

The discipline of constantly analyzing information, ideas and methods to assess reliability of a technology. Technology: the application of scientific knowledge usually in business & industry.

- Is a proven strategy for developing new technologies
- Generates a valuable source of contacts; People (designer, users, merchants, etc) linked to the product
- Generates case histories and learnings from successes and failures
- Index of developed reliability test methods for technologies
- The basis for RE-INVENTION

The Goal of Technology Assessment

"To produce dependable technology through the use of adequate design, materials and processes"

Define

Technology assessment is a conscious and organized effort to learn what went right and what went wrong when developing or controlling technology. It is an opportunity to ask, 'how can we make it better?'

Technology assessment is not against the idea 'if it's not broken, do not fix it', but is a way to understand why things worked and how can the success or failure be used on future improvements.

References and Resources:

Juran and Godfrey (1999) covers a wide range of topic including technology assessments.

Spiegelhalter (1998) writes on the topic Markov modeling and the Monte Carlo approach.

See Carlin and Louis (2009) for a treatment of Bayes methods for data analysis.

Ireson, Coombs, and Moss (1995), Lloyd and Lipow (1984), Klinger, Nadada, and Menendez (1990), and Doty (1989) include a range of discussion concerning technology assessments

A. 16 — Test Capacity

Concept

The basic assessment of how much product and to what parameters can we SAFELY test.

Normally identified in man hours for teams, cubic feet for environmental chambers, maximum weight for vibration and shock tables, plus the maximum stresses applicable (Hz, Voltage, Temperature, Humidity, Amperage, Lumens, G's, Kilograms, etc.) and how quickly can we transition product to the stress extremes.

Test Cycle Time is an obvious driver for test capacity as the shorter the test cycle, the more products we over can test with the same resources.

Define

The clear definition of test capacity is a key element of service for every test facility.

- Is necessary for the test facility as well as for its customers to have clear understanding of the real capacity limits (Human imposed and equipment imposed) in order to obtain test results that are both accurate and valid.
- Is helpful to document the rated capacity for every piece of equipment in the lab as well as for the team who sets up and monitors tests and evaluates results, to be able to clearly communicate the rated capacity as needed.

References and Resources:

Ireson, Coombs, and Moss (1995), Lloyd and Lipow (1984), Klinger, Nadada, and Menendez (1990), and Doty (1989) include a general discussion on testing capacity topics.

A. 17 — Test Planning

Concept

Test Planning: the first step necessary before any test activity is performed.

Aimed to answer the question; 'what do we need to know from this experiment and when?'

A comprehensive test plan describes the objective, approach, resources and schedule of intended testing activities; identifies test items, features to be tested, the testing tasks, who will do each task, and any risks requiring contingency planning.

Define

The following are elements necessary in a test plan:

- Provisions to make the test REPEATABLE (applicable Industry Standards?)
- Definition of failure
- Definition of success
- Necessary statistical significance
- Definition of sample sizes
- Sample preparation needed
- Equipment preparation needed including any consumables
- Test parameters
- Can an acceleration factor be used?
- Allocation of test capacity
- Time (how long will the test be? when do we need results? when can we run the test?)
- Who will perform the analysis and when will results analysis be run?
- Results reporting (who needs to be informed of the results? what format to use?)
- Who will assign corrective actions?
- Risk factors (identification and mitigation)

References and Resources:

See Whittaker (2003) for discussion on software testing.

Wiggins and McTighe (1998) for the concept of creating tests that provide meaningful results.

Ireson, Coombs, and Moss (1995), Lloyd and Lipow (1984), Klinger, Nadada, and Menendez (1990), and Doty (1989) include a range of discussion concerning testing and test planning.

A. 18 — Test Tool and Method Development

Concept

Tool and / or Method Development: the design and development activities that result from the need to test a new parameter or a new product that does not have a parallel in the industry

It is regrettably common that a product development effort misses the need to generate new tools (hardware or software) or new test methods necessary to verify functionality of new products. These "surprises" result in significant unexpected hits in schedule and budget; therefore one needs to be anticipated by a comprehensive test planning effort beforehand.

When developing new products or services it is imperative to analyze if the capacity to test exists.

Does it currently exist or if it will have to be developed.

The investigation of "industry standard" tools and methods, the design, generation and implementation of new ones, adds significant time and resources draw to any project.

When it is known that new tools and methods will be needed, the necessary activities, times and expenses need to be made part of the overall project tracking systems.

References and Resources:

See the EPA (2018) National Risk Management Research Laboratory site for information and programs concerning environmental related technologies. Project to develop tools and technologies to measure aerosol chemical composition and particle dispersion through space.

Consider Wiggins and McTighe (1998)

See Nielsen (1993) for usability concept discussion.

Ireson, Coombs, and Moss (1995), Lloyd and Lipow (1984), Klinger, Nadada, and Menendez (1990), and Doty (1989) include a range of discussions on tool development for reliability work.

A. 19 — Managing Reliability Budgets

Concept

A maximum Failure rate allocated for a component or a product. This target is normally defined to maintain a product or system needed reliability according to market needs.

As an example, in a market where products have a reliability of 5 years, each component of the product will have a maximum failure rate that supports a total product Reliability of 5 years or more when the failure rates of all components are added together.

In the same manner, when the Reliability of a system is a condition, each product that is part of the system will need to accommodate a maximum failure rate supporting this limit.

Example for a New Printer Reliability Budget

Build	MCBSV Target	MCBF Target	Page Volume Goal	MCBJ
BB1.1	6,300	9,450	300,000	1,000
BB2	11,500	17,250	500,000	2,000
LP1	22,700	34,050	2,500,000	4,000
LP1.1	35,600	53,400		6,000
LP1.2	42,000	63,000	2,500,000	8,000
LP2	46,000	69,000	3,000,000	9,000
LP2.1	50,000	75,000	3,000,000	10,000
PP	50,000	75,000	3,000,000	10,000

MCBSV: Mean Copies Between Service Visits

MCBF: Mean Copies Between (Hardware) Failure

MCBJ: Mean Copies Between James

During a new product introduction, it is common to set Reliability budgets for each phase of the project, allowing for Reliability Growth that supports the Reliability goals needed for the product release.

A. 20 — Communicate Risk

Concept

Methods and examples to convey magnitude and significance of uncertainty in reliability metrics, models and measures used in product decisions.

Risk

The possibility of suffering harm or loss of life. The communication of danger.

Risk Exists on Different Levels

1. On the level of Production: for the consumer, for the producer and for the schedule.
2. On the level of Analysis: for error of interpretation of field and test data generating erroneous corrective actions.

To mitigate risks, they should be acknowledged, estimated, managed and communicated effectively across the organization.

Scope

Is	Is Not
Expressed in potential costs exposure when possible	Simple formula or graphic
Common metrics and terms	One approved method
Both hard and soft metrics	One size fits all
List of methods to convey risk	
Directory of resources to help communicate risk	
Situation and audience dependant	

Advantages

- Provides input into design trade-off decisions
- Provides insight into areas of uncertainty
- Alerts need for preventative activities
- Attempts to look forward in time with clarity

References and Resources

Ireson, Coombs, and Moss (1995), Lloyd and Lipow (1984), Klinger, Nadada, and Menendez (1990), and Doty (1989) are good general references.

A. 21 — Data Strategy

Concept

A systematic plan of action focused on obtaining, securing, avoiding duplication and maximizing availability, distribution and overall value of data.

The overwhelming abundance of data in our days makes it imperative to consciously rationalize the way we generate, treat, and use and store information.

Define

Reliability estimations, verifications and product decisions are the result of the analysis of data.

It is imperative to have a conscious and defined management strategy for Reliability Data aimed to make sure that:

- Relevant data is easily obtainable
- Data is accurate and error free. Automatic data capture is always preferred
- Waste efforts to re-generate existing data are minimized
- Data is easily accessible

- Data is presented in the simplest format
- Data is efficiently distributed to ALL pertinent parties
- Data is securely stored and maintained

Data management Strategies should be evaluated and revised constantly, preferable according to periodic calendar.

This is a responsibility for Reliability Managers and Reliability Engineers.

References and Resources:

See Axson (2003) for a run down of best practices.

Consider Amendola and Keller (1987) for their focus on reliability databases.

Ireson, Coombs, and Moss (1995), Lloyd and Lipow (1984), Klinger, Nadada, and Menendez (1990), and Doty (1989) include general discussions around data and data handling.

A. 22— Cost of Reliability (or Un-Reliability)

Concept

The costs of low reliability go straight to the profit margins by reducing customer satisfaction, increasing warranty expenses, generating negative advertising directly from consumers and hampering our opportunities for sequential sales.

Warranty: the direct expense incurred when a customer

- returns a product
- requests a repair
- requests servicing
- calls for support

You can go broke fixing problems, but you can get rich preventing them. — Phil Crosby

Scope

Is	*Is Not*
Collection of potential and actual expenses that include costs of rework, scraps, returns, customer calls, additional sales efforts, etc.	Only warranty expense
Caused by product that is deemed unreliable by customer	Only material & scrap costs
May include brand goodwill	Recognized accounting practice
	Available through financial systems
	A fixed number or value
	How to change costs

Advantages

By consciously deciding to measure the cost of un-reliability, an organization can:
- Have a way to indirectly gauge customer satisfaction
- Drive design decisions based on customer satisfaction estimates

Resources and References:

Crosby (1998) explores the cost of non-conformance in detail.

Juran (1999) offers a range of online and in person training on lean six sigma topics.

NASA (2018) is a PDF document and presentation on the cost of quality for large programs.

On Paper

Ireson, Coombs, and Moss (1995), Lloyd and Lipow (1984), Klinger, Nadada, and Menendez (1990), and Doty (1989) include general discussions around the cost of reliability.

A. 23 — Reliability Curriculum

Concept

A resource for all people who need reliability skills in a specific area:
- New reliability engineers
- Experienced reliability engineers
- Reliability managers
- Other engineers and managers

Curriculum

a group of related courses, often in a special field of study.
(In this case, the special field is reliability)

Scope

Is	*Is Not*
Recommended and prioritized sets of reliability courses/training for different engineer/manager types	Taught by a reliability expert
Closely tied to the reliability foundation	Other disciplines
All the information needed to sign-up for the recommended courses	Computational modeling
Leveraged off of existing materials where possible	
Developed by reliability experts where needed	
Includes training for other reliability skills	

Is	*Is Not*
Expanded as course materials become available	
Promoted/Advertised for both reliability and other staff	

Advantages

- New reliability engineers can get up to speed faster
- Experienced reliability engineers can expand their knowledge
- Other engineers will learn how they affect reliability and what they can do to affect it positively
- Management will learn about reliability management best practices and understand how they can affect reliability positively

References and Resources

See ASQ (2018a) and ASQ (2018b) for details on the Certified Reliability Engineer (CRE) program including a copy of the CRE body of knowledge.

Visit Munk (2018) for a basic introduction to the six sigma approach.

See NIST (2018) for a compilation of statistical methods in an online handbook format.

Reliasoft (2018b) is a purveyor of reliability statistics software and provide training plus an informative newsletter and archive of articles.

See Reliability Center (2018) for root cause analysis resources, training, and consulting services.

See Weibull.com (2018b) for a compilation of resources, online text books, software, and tutorials)

On Paper

Ireson, Coombs, and Moss (1995), Lloyd and Lipow (1984), Klinger, Nadada, and Menendez (1990), and Doty (1989) include a wide range of reliability topics.

A. 24 — Failure Criteria

Concept

A set of guidelines to aid in the classification of results during reliability testing.

Criteria

A standard, a rule, or a test on which a judgment or decision can be based; a means for judging or assessing performance.

Scope

Is	*Is Not*
Set of parameters that reflect the minimum performance expected from the tested samples or systems.	Used to establish liability
Loss of ability of a sample or system to perform a previously specified function or to hold a defined characteristic.	Empirical
Established BEFORE any testing starts.	Variable from sample to sample or test to test for each particular set of functional parameters.
Agreed criterion between Design and Marketing disciplines.	

References and Resources

BMP (2018) include a range of manufacturing best practices including material on failure criteria definitions for electronic equipment.

Find failure criteria theories for machine parts and structural members online at efunda (2018).

Visit James (2018) for an article on fatigue failure as a design criterion on mechanical and structural design.

Pizzo (2018) outlines fracture mechanics tutorials available online and through San Jose State.

On Paper

Ireson, Coombs, and Moss (1995), Lloyd and Lipow (1984), Klinger, Nadada, and Menendez (1990), and Doty (1989) include a wide range of reliability topics.

A. 25 — Goal Setting

Concept

Goal setting is the basic process of defining a fair, achievable and ambitious target upon which to focus reliability.

> There is no achievement without goals — Robert J. McKain

Appropriate goal setting will define direction, focusing enterprise resources toward a quantifiable goal.

Effective goals have several common characteristics:

- Achievable
- Measurable
- Aggressive (ambitious)
- Documented
- Posted, clearly known and understood by all interested parties
- Set considering the available resources (time, money, materials, process, etc.)
- Concrete deadline established
- Tasks broken down with clearly defined owners

References and Resources:

See Mind Tools Content Team (2018) online for a general discussion on goal setting.

See Kassab (2018) for a short article with three goal setting tips.

On Paper

See Smith (1999) for general treatment of goal setting.

See Blanchard and Johnson (2015) for a general treatment of time management.

See Ziglar (2002) for a general treatment of goal setting.

Ireson, Coombs, and Moss (1995), Lloyd and Lipow (1984), Klinger, Nadada, and Menendez (1990), and Doty (1989) include brief discussions on goal setting.

A. 26 — Intellectual Property

Concept

IP (Intellectual Property) – a product of the intellect that has commercial value included copyrighted property such as patents, business methods, manufacturing methods, testing methods, industrial processes, etc.

Define

Any unique, innovative approach to do our jobs is in essence an invention and therefore manageable by current IP policies. We all (regardless of our field) have the opportunity to add to IP every time we solve a problem, it is necessary though that we ask ourselves a set of very basic questions:

- Is this approach unique?
- Does it offer any advantages over previous methods?
- How do I know this method works?

All new procedures, test methods, as well as devices need to be analyzed for potential intellectual property.

References and Resources

See Stim (2017) and Miller and Davis (2012) for additional details concerning intellectual property.

A. 27 — Metrics

Concept

Examples and guidelines for use of reliability metrics during program development.

Metric

a way of measuring performance.

References and Resources

Ireson, Coombs, and Moss (1995), Lloyd and Lipow (1984), Klinger, Nadada, and Menendez (1990), and Doty (1989) include brief discussions on metrics.

A. 28 — Use of ODM/OEM (Outsourced) Applications

Concept

Guidelines and resources to consider when establishing an Original Design and Manufacturer (ODM) or an Original Equipment Manufacturer (OEM) relationship.

ODM/OEM

Strategic partner which can provide specific end-to-end capabilities from design to support

Scope

Is	Is Not
Best practices guidelines	The only way
Tips for reliability goals	Central control of decisions
Contract templates	An approval process
Includes all phases of relationship	Product specific
	Contract or internal manufacturing

Advantages

- Reduce errors and mistakes in ODM/OEM relationship
- Leverage central and consistent approach
- Increase awareness of best practices for reliability
- Reduce product failure rate and warranty exposure
- Shorten time-to-market

Deliverables

Completion Criteria	Priority	Team Required
All links	Medium	Reliability Engineers
Directory documented for sources of information		Reliability Experts
Guidelines/strategy for OEM/ODM relationships		Others

Approach

Technology Required	Information Sources	Project Framework (WBS)
Web linkage	Web databases	Define types of risks
Document readers	Government databases	Define types of ODM/OEM relationships
	Reliability Engineering society databases	Identify databases guidelines
		Collect copies and sources
		Arrange links and copies into categories
		Put links and copies into categories
		Periodically review integrity of links and copies

A. 29 — Process Control Strategy

Concept

Process Control

The deliberate adjustment of a series of actions, changes, procedures or functions necessary to create a result, a service or a product.

Process Control Strategy

The systematic plan of action focused on company processes (marketing, design, supply chain, production, inventory, distribution, services, etc.) aiming to streamline, maximize efficiencies, minimize costs, while providing best in class customer service.

Define

Each organization in the company has a specific strategy to meet the requirements of the market it serves.

A good example is the strategy delineated for a manufacturing process. The characteristics of this strategy can be translated to any other process across the company:

- Will develop trace ability on every product so the entire manufacturing history is easily known for each product
- Every manufacturing process will post defined performance metrics including volumes produced, quality yields, resources used, waste generated and on time deliveries against targets
- Our ability to efficiently solve quality and yield issues will be measured against targets
- Will constantly retrieve failed units from the field to calibrate our quality standards, drive manufacturing processes and estimate customer satisfaction levels

Reliable innovation at a price our customers can afford, delivered with an experience that set us apart. We deliver high tech, low cost, best customer experience. — Carly Fiorina.

Reliability Note

Certain activities can be designed into the process to aid in the quantification and maintenance of Reliability estimations, like:

- Burn-in of initial production to assure the production processes are controlling the key elements of the product performance.
- Qualification testing for any new process.
- Ongoing Reliability Testing (ORT) is an essential testing of the manufacturing process. Refer to the reliability skill labeled "Product Qualification" for further details on ORT. Results from

ORT testing should drive production processes to fine tune the performance variables critical to the life of the product.

References and Resources:

See Anupindi, et. al. (2012) and Smith and Fingar (2012) for more on business processes.

See Lipták (1995) for a focus on instrumentation of processes.

See Wilson and Harsin (1998) for process documentation best practices.

A. 30 — Regulatory Compliance

Concept

The adherence to standards designed for a particular industry, aimed to ensure user safety and environmental friendliness of products or services offered in a specific geography.

Define

Regulatory compliance is typically achieved by adherence to a regulatory body that publishes parameters that must be met. In many cases, this regulatory body is a government agency or other authorized organization. In each case, requirements include the submission of a plan for compliance, submission of documentation proving adherence to the standards, audits performed by the regulatory body and reporting mechanisms to address and resolve issues. In many cases, the adherence to a minimum number of regional marks is a prerequisite for sales and distribution in a specific geography. In every case, product that meets regulatory compliance has the right and obligation to bear the symbol or mark of the regulatory body on each label or box of product. Product that fails to meet the requirements for a mark can be recalled and production stopped until compliance can once again be proven.

Reliability Note

Some of the regulatory marks DO specify performance parameters for equipment that relates directly to Reliability performance (like tolerance for ESD discharges, fast transient discharge, conducted emissions and more).

These requirements should be contemplated in an 'Ongoing Reliability Program' that is aimed to identify deviations in product performance that could jeopardize compliance to the marks.

An Ongoing Reliability Program (ORT) is a constant activity performed according to a calendar, where the frequency of tests and sample sizes are defined by the volumes of product produced. Samples should be pulled from stock, from every production facility associated with the product and performed to product failure (destructive testing) to identify product performance margins and any deviations from original DVT (Design Verification Tests) and Product qualification tests. Refer to the reliability skill – Product Qualification.

References and Resources

On the Web
Visit Raymond Corporation (2018) for a listing of international recycling laws and Blue Angle (2018) for details on the German eco-label.

Visit UL (2018), CSA Group (2018, and TÜV (2018) for details on these certification marks.

On Paper
See Sparrow (2000) for a general discussion on all things regulatory.
See Dennison (1995) for regulatory compliance reporting practices.
See the regulatory handbook by PricewaterhouseCoopers (2001).

A. 31 — Resource Strategy

Concept

A set of best practices for product teams to obtain a successful and efficient use of their Reliability resources. The elements are:

- Resources needed and why
- When resources typically used
- Who provides inputs of needs and what inputs typically look like
- Ways to find out where to get resources and who to contact

A resource is anything, person, action, etc., to which one turns for aid in time of need.

Scope

Is	Is Not
Data	Rules
Facilities	Decisions
People (project & support)	Tactical
Money	
Space	
Equipment	
Parts	
Materials	
Time	
Ideas	
Processes	
Specifications	
Data Management	

Advantages

Establishes a long range map with well thought out strategy to set a course for tactical planning

- Helps keep short term disruptions from derailing a long term focus
- Allows for integrating project needs over a time period which is longer than any one project
- Promotes good business decisions in terms of long range planning and smoothing of funding demands

Resource Considerations

Things	*Approach Taken*
Resources needed ▪ quantity ▪ type ▪ money	Testing style ▪ HALT ▪ test-to-spec ▪ stress
Processes ▪ setup/teardown ▪ operations ▪ maintenance/calibration ▪ quality assurance ▪ data analysis ▪ failure analysis	Technology tested ▪ software/hardware ▪ portable ▪ consumer ▪ medical, etc.
Equipment/Facilities ▪ capability & capacity ▪ space	Data management ▪ raw ▪ reports
Test Vehicles ▪ baseline ▪ references ▪ control & investigative samples	Documentation ▪ test plans ▪ equipment ▪ schedules ▪ finance

Things	*Approach Taken*
	Metrics • competitive and goals

References and Resources

See Wallnau (2004) for an article on software component reliability.

ABS Group (2018) offers a seminar on quantitative analysis techniques.

See HBM Prenscia (2018) for an article on a range of reliability tools and techniques.

See Reliasoft (2018a) for an article on improving system reliability through component improvement

On Paper

See Hargadon (2003) for a discussion on innovation within organizations.

See Abrashoff 2002 for a comprehensive discussion on resource management and leadership.

Ireson, Coombs, and Moss (1995), Lloyd and Lipow (1984), Klinger, Nadada, and Menendez (1990), and Doty (1989) include brief discussions on resource management.

A. 32 — Risk Awareness

Concept

The conscious acknowledgement of the possibility of loss or danger

Define

The essence of the Reliability profession is to identify, quantify and mitigate risks of failure in our products or services – Anonymous

When considering risks for product failure, be aware, be very aware – Anonymous

When calculating the expected life of a product, or identifying failure mechanisms, or using any other Reliability tool (FMEA, DOE, etc.), we need to be very clear that the transcendence of the activity depends on how effective are we in informing the meaning and implications of the information gathered, making the organization aware of the risks involved and the possible actions needed to mitigate these risks.

Risks should always be analyzed from two perspectives: The risk to the consumer and the risk to the producer:

- Producer risk (alpha) = Probability of rejecting good product
- Consumer risk (beta) = Probability of accepting bad product

Acceptable reliability levels, maximum failure rates, sample plans for testing and inspection, etc., are ALL derived from the acceptable levels of these two risks.

References and Resources

See Crouhy and Galai (2014) for a few essentials.

See DeMarco and Lister (2003) for a discussion on software project related risk management.

Ireson, Coombs, and Moss (1995), Klinger, Nadada, and Menendez (1990), and Doty (1989) include brief discussions on risk management topics.

A. 33 — Statistical Data Planning

Concept

A set of methods and tools for exploring, analyzing and drawing conclusions from data from a process or test.

Define

Many tools exist on statistical data analysis. A few are:

- Pareto, Control Charts (for process control)

- Histograms, Line, Bar and Pie Charts
- T-tests, Z-tests
- ANOVA (analysis of variances)
- Normal, Binomial, Chi-square, Poisson, Gaussian, Weibull distributions
- Linear, nonlinear, and canonical Regressions, correlation, curve fitting, least squares
- Bayesian
- Probability distribution functions
- Many more...

The "key for success" on statistical data analysis lies on selecting the "right" tool to clearly highlight the problem area.

To help in the selection of the tool to use is important to have a clear idea of what parameter is critical in each case. Do we want to show:

- Contribution of a particular variable in a group? (use Pareto)
- Track behavior of a variable over time? (use line charts, control charts)
- Estimate future behavior based on past performance? (use regression, correlation, curve fitting)
- Analyze a series of events in relation to their whole group or other group populations? (use ANOVA, distributions, Bayesian), etc.

References and Resources

See Smith, et al. (2011) for a compendium of statistical analysis resources.

See StatPac (2018) for a survey of software and tools for professional researchers.

On Paper

See Newton and Rudestam (1999) concerning data analysis.

See Der and Everitt (2009) for details when using SAS.

Consider Kachigan (1991) for an introduction to multivariate data analysis.

For design of experiments see Montgomery (2017)

Ireson, Coombs, and Moss (1995), Lloyd and Lipow (1984), Klinger, Nadada, and Menendez (1990), and Doty (1989) include brief discussions on statistical analysis.

A. 34 — Statistical Test Planning

Concept

The design of tests to deliberately focus or highlight results of a particular performance variable or suspected problem.

Define

Strategic test planning provides procedures for ensuring that:
- Appropriate testing is performed throughout the testing cycle ensuring maximization of test coverage.
- Test metrics are developed and delivered to gauge progress and help effectiveness of the testing effort.
- Tests are documented with objectives and completion criteria clearly defined.

Paperwork generated by the test is minimized through efficient test planning.

Strategic Test Planning helps conduct the testing effort in the most efficient and effective way possible. Therefore the documentation of test strategies should be performed as early in the development cycle as possible.

References and Resources

Visit element (2018) for a range of product testing services.
On Paper
See Juran and Godfrey (1999) for a few chapters on statistical based testing.
See Whittaker (2003) for software testing.

See Rubin and Chisnell (2008) for a focus on effective testing.

See Nelson (2004) for a comprehensive treatment of accelerated testing.

Ireson, Coombs, and Moss (1995), Lloyd and Lipow (1984), Klinger, Nadada, and Menendez (1990), and Doty (1989) include brief discussions on statistical analysis.

A. 35 — Failure Modes and Causes

Concept

A set of best practices and resources for all engineers who need to identify potential and actual failure modes and their causes.

Failure

the condition or fact of not achieving the desired end or ends; not meeting the pre-established pass/fail criteria; a cessation of proper functioning or performance

Scope

Is	Is Not
Tries to identify, both potential and actual, failure causes and effects as well as estimate severity and frequency	A warranty that all failure modes will be identified and prevented (even when it is the ideal objective)
Best Practices/Training for: • Failure Analysis • Physics of Failure • Root Cause Analysis • Fishbone Diagrams • Fault Tree Analysis	

Advantages

- A good understanding of failure causes is necessary to ensure test coverage is complete
- Early identification of failure causes can be used as feedback during design
- Archive of known failure modes and their causes will shorten discovery time when working on leveraged or similar products

References and Resources

Visit James (2018) for the failure analysis page from the Materials Engineering school

On Paper

Ireson, Coombs, and Moss (1995), Lloyd and Lipow (1984), Klinger, Nadada, and Menendez (1990), and Doty (1989) include brief discussions on a range of failure modes.

Bibliography

Abrashoff, D Michael. 2002. It's Your Ship : Management Techniques From the Best Damn Ship in the Navy. New York: Warner Books

ABS Group of Companies, Inc. Quantitative Analysis Techniques. https://www.abs-group.com/Training/Course-Catalog/Quantitative-Analysis-Techniques/ (accessed November 18, 2018)

AIAG. 2008. Potential Failure Mode and Effects Analysis (FMEA) : Reference Manual. 4th ed. AIAG

Amendola, Aniello and A Z Keller. 1987. Reliability Data Bases : Proceedings of the Ispra Course, Held at the Joint Research Center, Ispra, Italy, 21-25 October 1985, in Collaboration with EuReDatA. Dordrecht ; Boston: D. Reidel Pub. Co. ; Norwell, MA, U.S.A. : Sold and distributed in the U.S.A. and Canada by Kluwer Academic Publishers

Andrady, Anthony L., Halim S. Hamid, and Ayako Torikai. 2003. Effects of climate change and UV-B on materials. Photochemical & Photobiological Sciences 2 (1): 68-72.

ANSYS, Inc. Rapid 3D Design Exploration. https://www.ansys.com/products/3d-design (accessed November 12, 2018)

Anthony, Robert N and Vijay Govindarajan. 2003. Management Control Systems. 11th ed. Boston: McGraw-Hill/Irwin.

Anupindi, Ravi, Eitan Zemel, Jan A. Mieghem, and Sudhakar D. Deshmukh. 2012. Managing Business Process Flows : Principles of Operations Management. 3rd ed. Boston: Prentice Hall

Ashby, M F. 2017. Materials Selection in Mechanical Design. 5th ed. Amsterdam: Butterworth-Heinemann

ASQ. Certified Reliability Engineer. https://asq.org/cert/resource/pdf/certification/inserts/CRE-insert-2017.pdf (accessed November 17, 2018)

ASQ. Reliability Engineer Certification CRE. https://asq.org/cert/reliability-engineer (accessed November 17, 2018)

Autodesk, Inc. Autodesk NASTRAN IN-CAD. https://www.autodesk.com/products/nastran-in-cad/overview (accessed November 14, 2018)

Axson, David A J. 2003. Best Practices in Planning and Management Reporting : From Data to Decisions. Hoboken, N.J.: J. Wiley & Sons

Bing image search for "evaluating design margins". 2017. https://www.bing.com/images/search?q=evaluating+design+margins (Accessed November 14, 2017)

Bing image search for "problem solving skills". https://www.bing.com/images/search?q=problem%20solving%20skills (Accessed November 14, 2017)

Bing image search for "root cause diagram". https://www.bing.com/images/search?q=root+cause+diagram&qpvt=root+cause+diagram (Accessed Novembe 14,r 2017)

Blanchard, Ken and Spencer Johnson. 2015. The New One Minute Manager. New York: Harper Collins

Blankenhorn, James C and Inc SMT Plus. 1996. SMT High Density Design & DFM. [San Jose, Calif.]: SMT Plus, Inc

Blanks, Henry S. 1992. Reliability in Procurement and Use: From Specification to Replacement. Chichester ; New York: Wiley.

Blue Angel. Blue Angel | The German Ecolabel. https://www.blauer-engel.de/en (accessed November 20, 2018)

BMP. 2018. The BMP Program: Technical Expertise. http://www.bmp-coe.org/index.html. (accessed November 20, 2018)

Burgelman, Robert A., Steven C. Wheelwright, and Modesto A. Maidique. 2000. Strategic Management of Technology and Innovation. 3rd ed. New York: McGraw-Hill.

Cadence Design Systems, Inc. Cadence Specman Elite. https://www.cadence.com/content/cadence-www/global/en_US/home/tools/system-design-and-verification/simulation-and-test-bench-verification/incisive-specman-elite-testbench.html (accessed November 15, 2018)

Carlin, Bradley P, Thomas A Louis, Bayes and Empirical Bayes methods for data analysis Carlin. 2009. Bayesian Methods for Data Analysis. Boca Raton: CRC Press.

Cohen, Lou. 1995. Quality Function Deployment : How to Make QFD Work for You. Reading, Mass.: Addison-Wesley

CPSC (United States Consumer Product Safety Commission). 2016. Recall List. Accessed November 4, 2016. https://www.cpsc.gov/Recalls

Crosby, Philip. 1998. Calculating the Price of Nonconformance. Compact disc. Philip Crosby Associates. Available at http://shopping.na3.netsuite.com/s.nl/c.721998/it.A/id.35/.f

Crouhy, Michel and Dan Galai. 2014. The Essentials of Risk Management. 2nd ed. New York: McGraw-Hill Education.

CSA Group. About US. https://www.csagroup.org/about-csa-group/ (accessed November 20, 2018)

Dassault Systèmes SolidWorks Corporation. SolidWorks. https://www.solidworks.com (accessed November 12, 2018)

Davey, Andrew. 2003. Detail: Exceptional Japanese Product Design. Kempen: Te Neues Verlag GmbH

DeMarco, Tom and Timothy R Lister. 2003. Waltzing with Bears : Managing Risk on Software Projects. New York: Dorset House Pub.

Dennison, Mark S. 1995. Environmental Reporting, Recordkeeping, and Inspections: A Compliance Guide for Business and Industry. New York: Wiley.

Der, Geoff and Brian Everitt. 2009. A Handbook of Statistical Analyses Using SAS. Boca Raton: CRC Press.

Doty, Leonard A. 1989. Reliability for the Technologies. Milwaukee, Wisc.: New York, N.Y. : ASQC Quality Press, Industrial Press

efunda. Failure Criteria. http://www.efunda.com/formulae/solid_mechanics/failure_criteria/failure_criteria.cfm (accessed November 19, 2018)

element. We Are Element. https://www.element.com (accessed November 20, 2018)

EPA National Risk Management Research laboratory, Georgia Tech Institute and University of Illinois, https://www.epa.gov/aboutepa/about-national-risk-management-research-laboratory-nrmrl (accessed November 20, 2018)

Fern, David. 2002. "Defect Tracking and Management Process". Software Quality Professional 5(1).

Fowler, A C. 1997. Mathematical Models in the Applied Sciences. Cambridge ; New York, NY, USA: Cambridge University Press

Freemantle, David. 2002. How to Choose : Why Our Greatest Successes Are a Reflection of Our Small Everyday Choices. London: Prentice Hall Business

Hanlon, J.T. Thu. 1996. "The future of components for high reliability military and space applications". United States. doi:10.2172/231537.   https://www.osti.gov/servlets/purl/231537.

Hargadon, Andrew. 2003. How Breakthroughs Happen : The Surprising Truth About How Companies Innovate. Boston, Mass.: Harvard Business School Press

HBM Prenscia Inc. Design for Reliability: Overview of the Process and Applicable Techniques. ReliaSoft. https://www.reliasoft.com/resources/resource-center/design-for-reliability-overview-of-the-process-and-applicable-techniques (accessed November 16, 2018)

Intertek. 2018. HALT Testing & HASS Testing. http://www.intertek.com/performance-testing/halt-and-hass/ (Accessed January 12, 2018)

Ireson, William Grant, Clyde F Coombs, and Richard Y Moss. 1995. Handbook of ReliabilityEengineering AndMmanagement. New York: McGraw Hill

James, M Neil. Tutorials on Learning from Interactive Failure. https://www.fose1.plymouth.ac.uk/fatiguefracture/ (accessed November 19, 2018)

Juran, Joseph M and A Blanton Godfrey. 1999. Juran's Quality Handbook. 5th ed. New York: McGraw Hill

Kachigan, Sam Kash. 1991. Multivariate Statistical Analysis : A Conceptual Introduction. 2nd ed. New York: Radius Press

Kassab, Mackenzie. 3 Things to Know Before Setting a Goal. Harvard University. https://www.extension.harvard.edu/professional-development/blog/3-things-know-setting-goal (accessed November 20, 2018)

Klinger, David J, Yoshinao Nakada, Maria A Menendez, and Bell Bell AT&T. 1990. AT&T Reliability Manual. New York: Van Nostrand Reinhold

Lawrence, Eric D. 2017. "Vehicle Recalls Hit a Record 53 Million Last Year." USA Today, November 13. http://www.msn.com/en-us/money/companies/vehicle-recalls-hit-a-record-53-million-last-year/ar-BBEUAiu?ocid=UE09DHP

Lifecycle Insights. 2018. Agile Product Development Processes. http://www.lifecycleinsights.com/tech-guide/agile-product-development-processes/ (Accessed May 29, 2018)

Lipták, Béla G. 1995. Instrument Engineers' Handbook. 3rd ed. Oxford: Butterworth-Heinemann

Lloyd, David K and Myron Lipow. 1984. Reliability : Management, Methods, and Mathematics. 2nd ed. Milwaukee, Wisc.: The American Society for Quality Contro

Mentor, a Siemens Business. Vista Electronic System Level Design. https://www.mentor.com/esl/ (accessed November 14, 2018)

Miller, Arthur R and Michael H Davis. 2012. Intellectual Property : Patents, Trademarks, and Copyright in a Nutshell. 5th ed. St. Paul, MN: West Academic Publishing

Mind Tools Content Team. Personal Goal Setting. https://www.mindtools.com/page6.html (accessed November 20, 2018)

Modarres, Mohammad, Mark Kaminskiy, and Vasilly Krivtsov. 1999. Reliability Engineering and Risk Analysis: A Practical Guide. New York: CRC Press.

Montgomery, Douglas C. 2017. Design and Analysis of Experiments. 9th ed. Hoboken, NJ: Wiley

Mozur, Paul and Su-Hyun Lee. 2016. "Samsung to Recall 2.5 Million Galaxy Note 7s Over Battery Fires." New York Times .September 2. https://www.nytimes.com/2016/09/03/business/samsung-galaxy-note-battery.html

Munk, Jared. Six Sigma 101: The Basics. Six Sigma Daily. http://www.sixsigmadaily.com/six-sigma-101-the-basics/ (accessed November 15, 2018)

NAHB (National Association of Home Builders.) Construction Industry Recalls. https://www.nahb.org/en/research/legal-issues/construction-industry-recalls.aspx (Accessed January 1, 2018)

NASA. Cost and Quality Planing for Large NASA Programs. https://ntrs.nasa.gov/archive/nasa/casi.ntrs.nasa.gov/19920010191.pdf (accessed November 16, 2018)

Nelson, Wayne. 2004. Accelerated Testing : Statistical Models, Test Plans, and Data Analysis. New York: Wiley

Newton, Rae R and Kjell Erik Rudestam. 1999. Your Statistical Consultant : Answers to Your Data Analysis Questions. Thousand Oaks, Calif.: Sag

Nielsen, Jakob. 1993. Usability Engineering. London: Academic Press..

NIST/SEMATECH e-Handbook of Statistical Methods, http://www.itl.nist.gov/div898/handbook/ (accessed November 17, 2018)

Onodera K. 1997., "Effective techniques of FMEA at each life-cycle stage," Annual Reliability and Maintainability Symposium, Philadelphia, PA., 50-56.

Panagacos, Theodore. 2012. Ultimate Guide to Business Process Management : Everything You Need to Know and How to Apply It to Your Organization. Seattle: Create Space Independent Publishing Platform

Pecht, Michael and Abhijit Dasgupta. 1995. "Physics-of-Failure: An Approach to Reliable Product Development." Journal of the IES September Vol. 38, No. 5: 30-34.

Pizzo, Patrick P. Exploring Materials Engineering. San Jose State University. http://www-engr.sjsu.edu/WofMatE/FailureAnaly.htm (last revised January 2, 2018)

Poli, C. 2001. Design for Manufacturing : A Structured Approach. Boston: Butterworth-Heinemann

PricewaterhouseCoopers. 2001. Compliance Link: 1998-1999. PricewaterhouseCoopers Regulatory Handbook Series. Abingdon-on-Thames, England: Routledge

Raymond Corporation. Recycling Laws International. http://www.raymond.com/RaymondCommunicationsRecyclingLawsInternational.html (accessed November 20, 2018)

Reliability Center Inc. Root Cause Analysis Training, Consulting and Software Since 1985. https://www.reliability.com (accessed November 19, 2018)

ReliaSoft Corporation. 2018a. Component Reliability Importance. ReliaWiki. http://reliawiki.com/index.php/Reliability_Importance_and_Optimized_Reliability_Allocation_(Analytical)#Improving_Reliability (accessed November 18, 2018)

ReliaSoft Corporation. 2018b. Software and solutions for reliability and maintainability analysis. Https://www.reliasoft.com (accessed November 18, 2018)

Rubin, Jeffrey and Dana Chisnell. 2008. Handbook of Usability Testing : How to Plan, Design, and Conduct Effective Tests. 2nd ed. Indianapolis, IN: Wiley Pub

Safe Kids Worldwide ™. 2018. Product Recalls. https://www.safekids.org/product-recalls (Accessed January 1, 2018)

Scheiber, Stephen F. 2001. Building a Successful Board-test Strategy. 2nd ed. Boston: Newnes

Schenkelberg, Fred. 2014. "Design Approaches to Creating a Reliable Product." Accendo Reliability. https://accendoreliability.com/2-design-approaches-to-creating-a-reliable-product/ (Accessed January 28, 2018)

Smith, A.K., Ayanian, J.Z., Covinsky, K.E. et al. J GEN INTERN MED (2011) 26: 920. https://doi.org/10.1007/s11606-010-1621-5

Smith, Douglas K. 1999. Make Success Measurable! : A Mindbook-workbook for Setting Goals and Taking Action. New York: Wiley.

Sparrow, Malcolm K. 2000. The Regulatory Craft : Controlling Risks, Solving Problems, and Managing Compliance. Washington, D.C.: Brookings Institution Press.

Spiegelhalter, D J. 1998. Markov Chain Monte Carlo in Practice. Boca Raton, Fla.: Chapman & Hall.

StatPac Inc. Survey software and tools for professional reserchers. https://www.statpac.com/ (accessed November 20, 2018)

Stim, Richard. 2017. Patent, Copyright & Trademark : An Intellectual Property Desk Reference. Fifteenth ed. Berkeley: NOLO

TÜV. About Us. TÜV Rheinland. https://www.tuv.com/en/usa/about_us/about_us.html (accessed November 20, 2018)

UL. About UL. https://www.ul.com/aboutul/ (accessed November 20, 2018)

Ulrich, Karl T and Steven D Eppinger. 2004. Product Design and Development. Boston: McGraw-Hill/Irwin

Wallnau, Kurt. Software Component Certification: 10 Useful Distinctions (CMU/SEI-2004-TN-031). Pittsburgh, PA: Software Engineering Institute, Carnegie Mellon University, 2004. http://resources.sei.cmu.edu/library/asset-view.cfm?AssetID=6985 (Accessed Nov 3, 2018)

Weibull.com. 2018. Military Directives, Handbooks and Standards Related to Reliability. https://www.weibull.com/knowledge/mil-hdbk.htm (Accessed January 14, 2018)

Weibull.com. Reliability Engineering Resource Website. https://www.weibull.com (accessed November 19, 2018)

Whitcomb, R. and M. Rioux, 1994. "Failure modes and effects analysis (FMEA) system deployment in a semiconductor manufacturing environment," Proceedings of 1994 IEEE/SEMI Advanced Semiconductor Manufacturing Conference and Workshop (ASMC), Cambridge, MA. 136-139.

Whittaker, James A. 2003. How to Break Software : An Example-rich Explanation of How to Effectively Test Software That Anyone Can Understand and Use. Boston, MA: Addison-Wesley

Wiggins, Grant P and Jay McTighe. 1998. Understanding by Design. Alexandria, Va.: Association for Supervision and Curriculum Development

Wikipedia SPC 2017. "Statistical process control". Last edited June 17, 2018. https://en.wikipedia.org/wiki/Statistical_process_control (Accessed November 14, 2017)

Willborn, Walter W O and T. C. Edwin Cheng. 1994. Global Management of Quality Assurance Systems. New York: McGraw-Hill

Wilson, Ray W and Paul Harsin. 1998. Process Mastering : How to Establish and Document the Best Known Way to Do a Job. New York: Productivity Press

Winston, Wayne L and S Christian Albright. 1997. Practical Management Science : Spreadsheet Modeling and Applications. Belmont, CA: Duxbury Press

Ziglar, Zig. 2002. Goals: Setting and Achieving Them on Schedule. (vocal performance). New York: Simon & Schuster Audio/ Nightingale-Conat.

Arthur Ray Hart

After college, Mr. Hart worked for Navy Microelectronic labs in failure analysis, MRC consulting doing modeling of radiation effects on satellites, and then (over the last 34 years) HP Personal group in the area of product development and technology applications. At HP, product types included integrated circuits, calculators, inkjet printers, projectors, and TVs. These products were developed using the principles discussed in this book. He was referred to by his peers at HP as 'Mr. Reliability'. The experience in the Navy labs providing a perspective of things that can go wrong with the best design approaches to use. Education background was in Solid State Physics (MS) and Mathematics (BS) at Western Washington State University.

www.ingramcontent.com/pod-product-compliance
Lightning Source LLC
Chambersburg PA
CBHW071714210326

41597CB00017B/2471